ALGORITHMIC MATTER THEORY

PAULO FRAGA

ALGORITHMIC MATTER THEORY

FUNDAMENTALS FOR THE PHYSICS OF THE THIRD MILLENNIUM

Translated by
Jhullyana Martins Soares

Translation reviewed by
Andressa Velloso Pinto

May/2023

Cataloging-In-Publication (CIP)
(eDOC BRASIL)

F811a	Fraga, Paulo Fernando Vargas, 1962-. Algorithmic matter theory: fundamentals for the physics of the third millenium / Paulo Fernando Vargas Fraga; translation Jhullyana Martins Soares. – Curitiba, Brazil: Author's Edition, 2023. 142 p. : 15,6 x 23,3 cm Includes bibliography Original title: Teoria da matéria algorítmica ISBN 978-65-5872-492-6 1. Physics. 2. Science – Philosophy. 3. Scientific Theory. I. Title. DDC 530.1

Prepared by Maurício Amormino Júnior – CRB6/2422

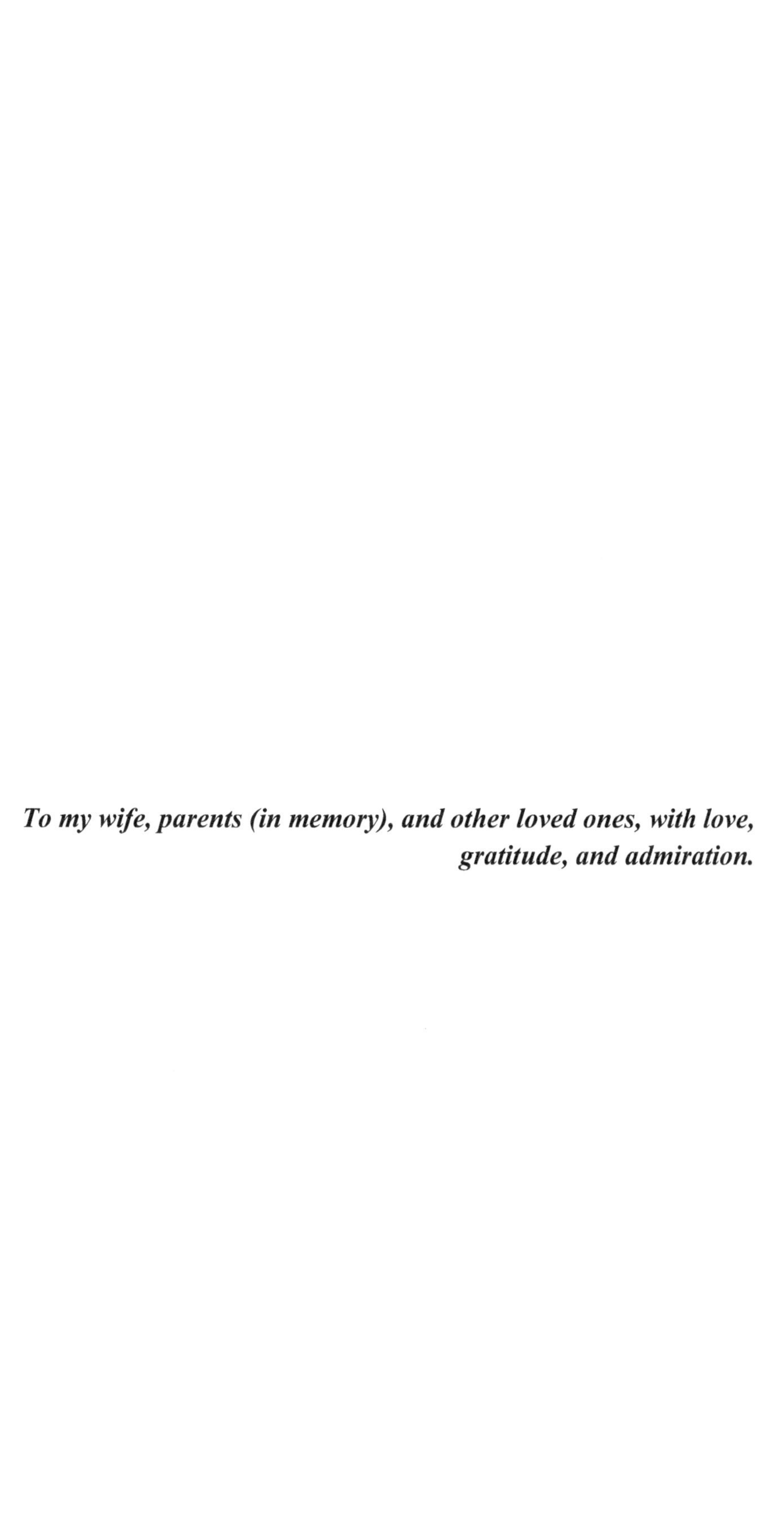

To my wife, parents (in memory), and other loved ones, with love, gratitude, and admiration.

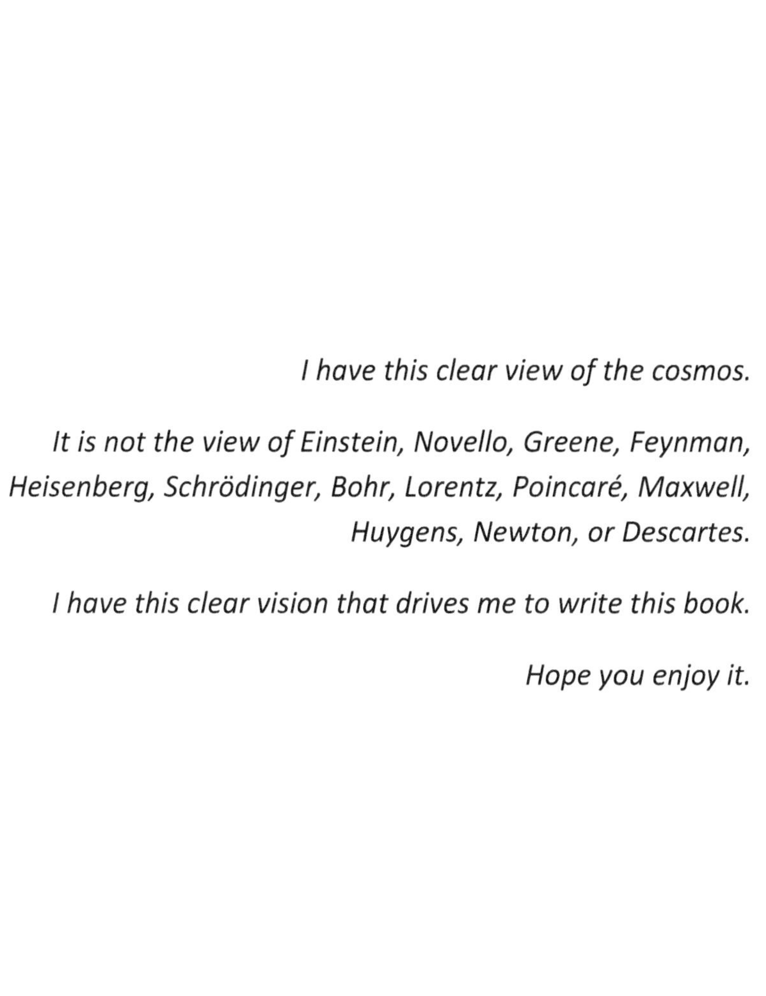

I have this clear view of the cosmos.

It is not the view of Einstein, Novello, Greene, Feynman, Heisenberg, Schrödinger, Bohr, Lorentz, Poincaré, Maxwell, Huygens, Newton, or Descartes.

I have this clear vision that drives me to write this book.

Hope you enjoy it.

CHAPTER I

THE SPACE-TIME CRITIQUE

Revolution of physics in the 20th century

At the beginning of the last century, Albert Einstein became one of the main exponents of a scientific revolution that helped shape the technological society in which we live today. Several scientists tried to replace some Newtonian concepts with relativistic ones, but little compares to the rupture provoked by Einstein in fundamental concepts such as gravitation, space, time, and their interactions with matter and energy.

In 1905, Einstein presented the theory that became known as Special or Restricted Relativity, based on works by Maxwell, Fitzgerald and Lorentz, having been largely preceded by Poincaré, although it is debated whether the work of this extraordinary philosopher, physicist, and French mathematician influenced Einstein directly or not. Perhaps Einstein's great differential concerning Poincaré in the paternity of Special Relativity was the conviction with which the German theorist considered unnecessary the existence of any kind of ether as an absolute reference for the explanation of fundamental physical phenomena, breaking with the idea adopted by the theorists who preceded him at the turn for the twentieth century.

Not satisfied with the still limited scope of Special Relativity, Einstein published General Relativity in 1915, a theory according to which matter distorts space-time, a non-granulated substrate whose curvature determines the gravitational effect, altering the rhythm of time and influences the trajectory of material objects and light.

Einstein's two theories of relativity have not yet been overcome as paradigms in the study of movement and gravitation on macroscopic scales.

Einstein also participated, following Max Planck, in the emergence of Quantum Mechanics, a fundamental physical theory for the description of phenomena that take place on ultramicroscopic scales, which received decisive contributions from Niels Bohr, Erwin Schrödinger, Werner Heisenberg, Paul Dirac, Wolfgang Pauli, Frank de Broglie, David Bohm,

Max Born, Richard Feynman and a whole legion of scientists who have sought to develop it up to the present day.

At the beginning of the 21st century, the Holy Grail of modern physics seems to be in the formulation of a theory capable of reconciling, with the elegance possible, two widely successful and deeply contradictory theories: General Relativity, which defines the gravitational phenomenon as an interaction between objects with mass and the fabric of space-time proposed by Einstein, and Quantum Mechanics, developed to explain and explore the microscopic world of atomic and subatomic particles.

Einstein himself never stopped believing in the existence of a unifying explanation for the identified discrepancies in the way matter and space appear to behave when observed at the macro and microscopic levels. Such an explanation should emerge from an as-yet unestablished conceptual model that scientists have called the Theory of Everything. String Theory developers (or Super String) believe they are on the way to finding this conciliatory model and claim the potential title of Theory of Everything for the set of its versions, known as M Theory, still in development.

The young Einstein

Let's imagine that Einstein was in his early twenties in today's world and that General Relativity, as it is known today, had been formalized by some other great scientist at the beginning of the last century. In such a hypothesis, the currently young Einstein would have no connection with the creation of that theory and would enjoy complete freedom to review all its premises, in the light of theoretical advances and new clues that scientific research has produced thanks to today's technology. It would be very difficult to assume that young and restless scientist would remain inert in the face of the information we have today, such as those that led to the almost unanimous acceptance of something like dark matter. Possibly, based on these new clues, the young Einstein would have been one of the great questioners of the fundamental assumptions established by General Relativity and Quantum Mechanics. Against the latter, the real version of the great scientist once rose up, despite appearing as one of its founders, with the famous phrase "I do not believe that God plays dice with the universe" – the exact words were probably not these –, and considered Quantum Mechanics an incomplete theory.

Among the new clues, perhaps the most important is the very discovery of the possibility of the existence of the so-called dark matter. The mere acceptance of this hypothesis by the scientific community leads to the need to promote a general revision of the main assumptions of modern physics. It obliges all interested parts, in the absence of the great German master, whether philosophers, physicists or mathematicians, professionals or not, students and other interested parts, to ask for permission and to look hard for any limitation in the currently accepted formulations, to reassess new and old ideas and, if necessary, look for something new somewhere in the past, trying to find any possible deviation from the route that has resulted in the scientific impasse of our days.

The contradiction between General Relativity and Quantum Mechanics plunges modern physics into a conceptual crisis that drags on for decades. This type of crisis, in science, usually favors the emergence of a new model that can reorient scientific production, without the constraints resulting from the exhaustion of previous models. This does not mean failing to recognize the merits of more remote or recent advances in knowledge, but seeking to identify its limits and eventually question some of its most enduring foundations, as occurred at the beginning of the last century.

A new theory of gravity and time

A simple explanatory theory, describing a reasonably elegant universe, is what I believe I can present at this time, a fact that imposes the important task of asserting, with the deepest conviction, the absolute absence of necessity the real existence of a continuous substratum such as space-time, as conceived by Einstein, to explain the gravitational effect and the relativity of time. The possibility of substituting space-time for a concept with greater explanatory power of physical reality makes the continuum conceived by Einstein an approximate image of reality, which has played an extraordinary role in the evolution of science's representation of the real world. I do not doubt that the effective replacement of the Einsteinian model with the model I am about to propose will open the doors for extremely more consistent, comprehensive and deep explanations of the most varied phenomena of the natural world, in all the scales we can imagine.

Is space-time real?

Let us take a look at the Einsteinian concept of the space-time continuum. It is time to challenge the acceptance of its ontological nature, from a new theoretical perspective, which represents the evolution of an ancient principle, going back at least to Descartes. We can call this principle the algorithmic principle. Before proceeding, I must say that the theoretical premise that I will adopt also implies the necessary negation of the uncertainty principle, one of the main pillars of the current quantum conception of the universe.

The uncertainty principle and the spacetime continuum are categories that are so strongly accepted and influential today that their definitive overcoming requires a theoretical articulation as comprehensive as it is consistent, a new and powerful principled scheme, both corroborable and refutable, capable of highlighting the need to attribute to those categories the condition of important scientific dogmas of our time.

The concept of the space-time continuum, for example, influences practically all current conceptions of how the universe works, including Quantum Mechanics and String Theory. The concept says that the universe is formed by a continuous "space-time fabric", which bends or deforms in the presence of matter (or energy), resulting from this curvature, the entire gravitational effect, which implies a distortion both in space and in time, magnitudes inseparably imprinted on the continuous substratum (space-time) that forms the emptiest part of the universe, the spatial vacuum where everything is immersed.

The idea of the space-time fabric is related to the notion of field, introduced by Faraday[1] and developed mathematically by Maxwell, who systematized the fundamental equations of electromagnetism. The notion of field emerged as a graphic representation of the lines of force associated with electric and magnetic charges and evolved from the notion of a mere resource of representation to a physical concept[2], which

---

[1] HAWKING, MLODINOW, 2011, p. 66

informs the behavior of space, that is, of the space vacuum, its physical capacity to react to the presence of electric charge in movement and to transmit electromagnetic waves[3], in a region of unlimited range (field), where the potential effects of this moving charge are manifested, decreasingly as the distance from the charge increases, until its effects become minimal and negligible. Similarly, the spatial vacuum reacts to the presence of matter and the mass associated with matter. So, in addition to the electromagnetic field, there would be a gravitational field sharing every portion of the void in the universe. Nowadays, physics seems to regard these fields as overlapping and independent, sharing the cosmic fabric with other fields as well, such as the Higgs field, so that the electromagnetic field reacts to charged particles or electromagnetic waves, the gravitational field reacts to mass and the Higgs field reacts to matter. Superimposed and independent fields. We'll see that Einstein didn't like that...

Space-time signifies an important philosophical turning point in physics. Space ceases to be an inert, Newtonian stage, where physical phenomena occur, and starts to assume an active role, interacting with matter and energy (currently conceived as a state of matter), deforming itself in its presence, influencing, in turn, the translative movement of material objects and also of light.

Two main reasons seem to have seduced Einstein to this inflection. The electromagnetic field equations, systematized by James Clerk Maxwell, which in Einstein's view seem to explain the propagation of electromagnetism without the need for the existence of a particulate medium, the supposed ether, and the absence of a satisfactory conception associated with this hypothetical medium, the existence of which was defended by several scientists, each in their own way, such as Christiaan Huygens' ether, imagined for explaining the wave nature of light. The famous Michelson-Morley experiment helped to bury in the scientific community, without this being the original intention of those two experimentalists, the hypothesis of the existence of Huygens' luminiferous ether, which was opposed to the corpuscular theory of light, defended by Isaac Newton. The electromagnetic ether - unlike Huygens's

---

[2] EINSTEIN, INFELD, 2008, pp. 109-131

[3] EINSTEIN, INFELD, 2008, p. 131

luminiferous ether - over which Maxwell himself built his successful electromagnetism theory, also lost strength in the early twentieth century, especially from the rise of Einstein's models, although Einstein himself argues that ether models have lost strength gradually due to the rise of the concept of an electromagnetic field and the difficulty of building an acceptable operation model.

> *Our attempts to discover the properties of the ether have led to difficulties and contradictions. After these bad experiences, it is time to forget about the ether completely and never try to mention its name. We will say: our space has the physical property of transmitting waves, thus omitting the use of a word that we decided to avoid.* [4]

As the conceptions of the ether that emerged until that point were not considered consistent with the results of the Michelson-Morley experiment, despite the defense presented by Fitzgerald, and endorsed by Lorentz, based on the extraordinary idea of the contraction of distance and time as the speeds increase, Einstein found reasons to propose the complete disregard of a cosmic medium formed by particles and made a remarkable effort to literally banish the use of the word ether in physics. Using arguments based on a very attractive logic, he extended the application of the concept of field to the gravitational phenomenon, with the proposition of a space-time continuum with four dimensions, consistent with Riemann's rigorous mathematical basis for curved geometric structures.

It was the beginning of the reign of General Relativity, the most recent speculative consequences of which include the possibility that space-time not only has four dimensions but, who knows, more than a dozen hidden and tiny dimensions, mathematically justified, as suggested by String theory. Another consequence of the space-time continuum concept is the basic notion in Quantum Mechanics, derived from Heisenberg's uncertainty principle, that the space-time tissue undergoes microscopic energy oscillations responsible for the constant creation and annihilation of matter and antimatter particles that almost instantly destroy themselves in pairs, in the so-called quantum frenzy, a theoretical

---

[4] EINSTEIN, INFELD, 2008, p. 149

phenomenon also called quantum foam. The quantum frenzy starts from the premise of the existence of the space-time continuum, but it doesn't seem at all incompatible with a grainy cosmic medium.

In any case, the dominant modern physics is built on the foundation of the space-time continuum, a physical-mathematical entity that interacts with matter, considered a condensed state of energy, such tissue being able not only to be shaped by matter but also to propagate it inside and, in the current quantum conception, randomly condense the energy and transform it into pairs of particles of matter and antimatter, which soon after collide, disintegrate and release that energy that was condensed. Constant transformations of portions of space-time into energy, energy into matter, and matter back into energy follow, stabilizing space-time again until a new random creation of particles by spontaneous generation of energy. I probably have not done enough research to realize whether there is any convincing explanatory proposition regarding how space-time, which is, in essence, a geometric fabric, creates non-condensed energy and transforms it into particles of matter, an assumption widely accepted by Quantum physics theorists and academia in general.

Returning to the macrocosm, it seems undeniable, within certain parameters, the existence of an extraordinary descriptive relationship between the curved geometry of General Relativity and the observable gravitational phenomenon. Einstein deduced that the existence of this positive descriptive relationship would be enough to assume the fundamentally geometric, mathematical, and physical nature of space itself, associated with time. This means that if you asked Einstein what the physical nature of space-time would be, he would answer that the physical nature of space-time is a geometric, mathematical nature. Let's say it again. For Einstein, space has the nature of an essentially mathematical substrate, although physical. For today's physicists, this is peacefully accepted as reality!

Alternative to space-time

Many people must have asked themselves whether the geometric nature is the only fundamental nature that can be attributed to space, given the successful relationship between the geometry of the curved four-dimensional space of General Relativity and the observable effects of gravitation. My deep conviction forces me to affirm that this inquiry is the key to the formation and acceptance of a unified theory that can open the doors to reach an unprecedented level in the understanding not only of gravity but also of the nature of matter, energy, mass, inertia, electromagnetism, time, light and possibly all the physical phenomena that reality and our capabilities have allowed us to observe and identify. This ambition is not particularly mine, but of humanity as a conscious species, that of walking in search of knowledge of the most intimate nature of our surroundings and the cosmic confines.

Brief mention of the proposal of a new theoretical model.

The model of reality that I am about to present does not emerge from any mathematical effort, but from inviting and deep descriptive gaps offered by current models, crying out to be filled, explored, and questioned. I have been calling this alternative model Algorithmic Matter Theory (AMT), or simply Algorithmic Theory (AT). If the current human civilization survives long enough for the few premises of the theory I undertake to present to be accepted by the scientific community, I think that this new scheme of representing the world will be able to guide the creation of a mathematics of its own, broad, and increasingly more complete, independent of the equations of General Relativity and the probability waves of Quantum Mechanics. I believe that this new theoretical scheme may eventually help to explain the reasons for the historical advances and deviations from reality inherent in those two main physical theories that are dominant today.

Mathematics and Physics

Perhaps the most revolutionary aspect of AMT is to once again accept and propose a space with only three dimensions. An irrepressible restlessness leads to the question: what if Einstein's curved space-time is an illusion induced by the incredible descriptive power of mathematics, which fails to capture the true physical nature of the universe? An illustration of this possibility is in order. Let's say a hypothetical alien civilization has a completely unique relationship to atomic matter than ours. Although its mathematics is reasonably developed, its members have no idea what a cannonball is. As much as they may be attacked by spiritually primitive war cultures, they, fortunately, do not suffer the effects of ballistic weapons that we consider conventional. Not because they have efficient shields or strong constructions, but because they hardly feel solid matter, since they are gaseous beings, despite living, circumstantially, above the rocky surface of a planet. Though immune to gunfire, they are intellectually intrigued because they somehow manage to track (or sense) the shots and determine the time until the projectile hits the ground, as well as the trajectory it describes. For them, even the concept of soil is very vague, dealing only with a specific region of their planet, perhaps a little more structured and difficult to penetrate, since they do not perceive solid things in the same way as we perceive them. They do not consider the projectile as a potentially harmful solid body but detect its movement by the effects it causes on its surroundings. For lack of necessity, they also didn't exactly develop a notion of falling, they only realized that the projectile ends its movement when it reaches that region of the planet (the rocky soil). If they are unable to abstract themselves from their condition as fluid beings and are incapable of conceiving the concept of a solid thing more deeply, despite being mathematically advanced and capable of deducing ballistic equations, they may simply be attracted by the idea that they are not being attacked by cannonballs, but experiencing a purely mathematical event, an attack of parabolas, described by quadratic functions.

Space and the Ether

Of course, the illustration may seem extremely crude, but it serves the purpose of helping to get the point across. Exploring the metaphor, a little further, we might ask whether the "fabric" of curved space-time, which would include time as a fourth dimension, could actually be composed, on an almost unimaginably small dimensional scale, by some unknown type of particles? Couldn't these particles act in the presence of material bodies endowed with mass so that their behavior would result in what is considered the curvature of space-time? Naturally, the attentive reader will argue that if the fabric of space-time is composed of particles, even extremely tiny ones, it ceases to be a continuum, if we consider particles as something that exists independently of the concept of excitation of any field. Except for this last hypothesis, the supposed continuous tissue becomes a medium composed of particles, a vast infinite ocean, or the limits of which are unknown. So, this medium, capable of being affected by matter, would be nothing more than a type of ether! This is precisely the point I want to get to. We know that Einstein condemned this word to the dungeons of the past, excising it from the dominant contemporary physics dictionary, despite renewed resistance from some thinkers. No matter how genial and admirable a man may be, he will never be able to inhibit eternally, absolutely, our thoughts about something that still admits to controversy. Einstein himself, for example, resurrected an idea of Newton's that was already largely dead and buried: the corpuscular nature of light.

We took an important step. We are admitting – after more than a century, although we are not the only ones – just by hypothesis, the existence of something like the ether. The reader friend must be at least curious by now to see where this is going since I can assume that you have not abandoned reading until this page. If we are going to continue, the next step is to ask what characteristics the ether must possess to accommodate the physical phenomena already proven or at least emphatically corroborated. In order not to disturb the sleep of the great master of Relativity, we could adopt a more up-to-date name for e…r, as he wrote when he mocked its existence and decreed its banishment from physics.[5] We could call it dark matter or sustaining matter. If your heart, dear reader, is beating fast when you realize what ground we're treading on, I suggest you stop a little, take a deep breath and decide if it's worth continuing. Are you ready? Can we continue? As for the possible disturbance of Einstein's sleep, if his rest can be really disturbed, it is most likely that he will have a good laugh at our insolence.

Space, according to our proposal, once again shelters an ultramicroscopic texture in its three dimensions, formed by particles physically distinct from space itself, returning to it the old Newtonian role of an inert stage, leaving the material action, now, again, in charge of a cosmic medium, as already proposed and widely accepted until a time that dates back to little more than a century. Thus, these particles do not belong to the structure of space itself, they just occupy it, forming a

---

[5] EINSTEIN, INFELD, 2008, p. 149

medium that extends, possibly, throughout the known universe and beyond. This does not imply that the particulate medium we postulate has already been rejected by physics as Newton's gravitational ether, Huygens' luminiferous ether and Maxwell's electromagnetic ether.

Recent conceptions of modern physics

Note that modern physics is also maturing this idea when it surrenders to the Higgs field and the existence of dark matter. Mario Novello speculated that the universe may be populated by 10^{120} gravitons, taking inspiration from Einstein's universal constant.[6] If the scientific community admits the existence of a medium of gravitons or the Higgs field, which would extend throughout the universe, and recommends the construction of an accelerator like the one of CERN, to detect the particles that inhabit this field, and if all known matter and energy flow in the universe, modern physics is admitting that there may be an ocean of particles through which matter moves and through which all electromagnetic waves propagate, such as radiated heat, visible light and gamma rays. Dark matter itself, so fundamental in the gravitational structure of galaxies, as it is believed today, if it is present between celestial bodies and in the interstices existing between their subatomic particles, is *necessarily* crossed by light. If dark matter, the Higgs field or a supposed graviton field exist and extend throughout the universe, there is no guarantee that light can – as claimed by Einstein – propagate itself in the absolute geometric vacuum. In the same line of reasoning, the electric and magnetic fields are also not guaranteed the ability to manifest themselves in a vacuum absolutely devoid of particles of any kind. The introduction of the theory of quantum gravity makes the analysis we are trying to develop a little more complex because apparently, that theory considers the particle of matter as a simple excitation of a field, that is, the particle does not exist independently of that field. According to this theory, the same would apply to the nature of all fundamental particles, conceived as excitations of one of the specific fields that overlap each other throughout the universe. Admitting this conception, inevitable gaps would need to be overcome. How do the fields overlap in space-time without interacting with each other, or how do they interact with each

---

[6] https://www1.folha.uol.com.br/folha/ciencia/ult306u13191.shtml

other? How do the excitations in the fields arise to form particles? And many other questions that quantum gravitation has not yet answered. In addition, such a model is essentially mathematical and as such is only a representation of reality and should not be confused with it.

The following criticism applies to the set of ideas that considers a fundamental particle of matter as an entity independent of the space-time continuum and any imagined field.

Continuity and permeability

One of the fundamental problems that we identified in the Einsteinian concept of space-time fabric is that it represents something simultaneously *continuous* and *permeable* to matter and energy. This brings some philosophical complications. Even if all the mathematics in the world is gathered together, it is very difficult to admit a physical structure that is both continuous and permeable, at least in the manner proposed by General Relativity. This skepticism does not just stem from the fact that the concept is contrary to normal human intuition. It is, rather, a simple problem of logic. Saying that something physical, which coexists and interacts with matter and energy, is absolutely continuous is equivalent to saying, in principle, that there are no intervals in its structure, except those occupied by matter or energy itself. So, in the first approach, for something to pass through the structure of space-time, it has to break through it. When ruptured, the structure automatically ceases to be continuous, at least there, in the region where the rupture occurred. In this approach, physical permeability implies physical discontinuity, as a simple logical and semantic result, if we consider continuity in a rigid denotation, that is, in a sense that does not admit any form of discontinuity.

There is, however, in a less rigid sense, at least one possibility of imagining a space-time as something continuous and permeable. If this medium is malleable enough, it could encapsulate something passing through it, as an air bubble is encapsulated by the surrounding ocean water as it moves toward the surface. This possibility applied to space-time would require very sophisticated dynamics for a non-granular medium. On the other hand, in media formed by particles, such as air and liquid water, this type of dynamic is very common in nature.

It could be argued that a material body such as a planet, when traversing space-time is not properly encapsulated by it, as the matter of the planet is not continuous. There would thus be a mutual permeability between the planet and the space-time substance. But assuming that matter is formed by corporeal particles and not mere excitations of a field, the encapsulation would be microscopic. In any case, the phenomenon of displacement of an encapsulated material particle would require, if not a constant rupture of the space-time fabric, at least a complex dynamic of displacement of each capsule – a portion of an absence of space-time – that is formed around every particle of matter in motion, such as quarks, electrons and photons. Furthermore, a surface would be formed in the continuum, in the region of the capsule shaped by the particle. There would also be an obvious slippage between the surface of space-time and the surface of the material particle. Space-time would be displaced by the particle, as a submarine does with water.

Continuity, permeability, deformation, compression and elasticity

In the relativistic model, space-time can be curved by matter. In this case, for being a substance, understood as a physical entity with independent existence, like matter and the field – the electromagnetic field, for example –, space-time must be compressible, capable of being locally deformed intensely. Let's consider, for simplicity, only the local deformation, without forgetting that the curvature caused by matter propagates infinitely – as it is inversely proportional to the square of the distance, the intensity of the curvature decreases a lot with increasing distance, but tends to manifest itself until the infinite. In addition to being able to be curved locally by matter, intensely, space-time must be sufficiently elastic to re-occupy three-dimensional space (or four-dimensional space-time) previously occupied by the moving matter particle, the same occurring on large scales. An extra note is in order here: in logical terms, the simple fact that it can be curved by matter makes space-time a privileged reference for the movement of any material body, regardless of the human capacity to deal with a reference of this nature.

The curvature caused by a body in motion adheres to the displacement of this body. Therefore, body and curvature move in relation to space-time itself as a whole. Defenders of Einsteinian mathematical space-time will probably say that I am using Newtonian language for a non-Newtonian object. Yes, that may be true. However, it becomes imperative to consider that celestial bodies are always Newtonian, as well as some of their attributes or states such as mass, velocity, acceleration, diameter, hardness, and inertia. If the exclusively mathematical space-time intends to physically interact with these material bodies, we cannot help but speculate about the nature and form of this interaction between the geometric substance and the Newtonian substance. There has to be an interface between these two worlds that allow this interaction. Either the mathematical substance has to acquire some Newtonian properties – deformation induced by material bodies, for example – or the Newtonian substance has to take on non-physical properties in the Newtonian sense, such as reacting to a strictly mathematical command of space-time. This last hypothesis would not solve the problem because these non-physical properties at some point would have to interact with the physical properties of the same material body, be it a corpuscular particle with a certain extension, in the hypothesis analyzed here, be it a rocky planet or a star. There is no way to keep the two worlds isolated from each other and at the same time admit that they are molded by each other's presence. In this scenario, we can try to explore the logical path that seems most promising, that of adding some Newtonian properties to mathematical space-time, so that this is not a ghostly entity that determines the fates of matter without any kind of concrete exchange.

To state that we cannot speculate about the hypothetical physical mechanism by which space-time transmits information to matter, and vice versa, at a macro or microscopic level, because Newtonian language is not adequate for space-time, is perhaps equivalent to elevating mathematics to the category of divinity, designs of which cannot be questioned by human reasoning. I believe that we will overcome the argument of this supposed inadequacy of language. If matter and space-time interact, as proposed by Einstein, science must try to answer how the microscopic mechanism of this interaction works and how it results in the famous macroscopic deformation of space-time in the presence of a material body composed of particles endowed with frenetic movement, as its vibration due to heat, and of translative movement with the material body they compose. This question is fundamental for the development of science.

The microscopic genesis of deformation of the space-time continuum

Returning to the speculative exercise, we are faced with a Newtonian corporeal particle, like an electron before current quantum models, just to simplify the exercise, revolving around the nucleus of an atom. On this size scale, disregarding the metaphysical premise of mathematical primacy in the genesis of physical phenomena, a new question arises. What could be the extent of the microscopic compression caused in space-time by the corpuscular particles of matter (still from a Newtonian perspective, as a mere starting point), located in the most internal regions of the large body? So that these particles can maintain a relatively stable average distance between themselves, in the structure of the atoms of the material body, considering the natural agitation of the particles, there must be constant compression and decompression of space-time in the interstices between the elementary particles of this body, given the premise that matter deforms space-time. We can imagine that the deformation must involve the effect of stretching one portion of space-time, contracting another, by the effect of each particle of matter in constant agitation. How to build a descriptive bridge between this individual microscopic effect of the material particle and the deformation that the macroscopic material body as a whole supposedly causes in the external region, as postulated by General Relativity? As far as I could deepen the study, Einstein did not enter into this merit, but the reasoning developed so far points to a combination between compression and decompression caused by the displacement of matter particles, with very little potential effect in the region external to the surface of the body, attributable only to the small deformations of space-time caused by the displacement of the most superficial corporeal particles of the body. In the relativistic model, it becomes extremely difficult, if not impossible, to compose a dynamic that explains the cumulative deformation that the set of material particles of a massive body can cause in the continuous physical-mathematical medium that surrounds it at distances much greater than the material extension of the body itself, such as the noticeable gravitational effects of the Sun throughout the Solar System, for example. This may make unfeasible the idea of the microscopic genesis of the large gravitational deformation caused by particles of a massive object, responsible for generating such noticeable gravitational effects, at distances thousands of times greater than the object's diameter. In other words, as material bodies of large dimensions are discontinuous, how can we conceive that the large deformation in space-time that they cause is the result of the sum of the microscopic deformations caused by each of their particles individually? And how to conceive that it is not?

The natural reaction to these considerations, as we have already said, would be to claim that they are supported by reasoning shaped by classical mechanics, in aspects already refuted by Relativity and by Quantum Mechanics itself. We can counter-argue again, remembering that General Relativity does not refute the fact that matter is formed by microscopic particles that keep immense interstices between them. In this case, these particles and the interstices need to be considered in the interaction between matter and space-time. Even if a material body could cause an amplified deformation effect in a non-Newtonian physical medium, that is, of an essentially mathematical and, nevertheless, physical nature, the difficulty of explaining the mechanism of concrete interaction between the four-dimensional curves of this geometric medium and the elementary particles that constitute material bodies would remain. The quantum gravity model tries to provide an answer to this difficulty, but because it is also an exclusively mathematical description, it equally touches on the problem of the relationship between mathematics and physics, already outlined earlier and which will be discussed in other ways later in this book.

We can imagine that the space-time curved by an immense star alters, due to the deformation of the lines that microscopically represent the continuum, the relationship of distances between the material particles inside a small body that moves in the vicinity of the large star. We could consider an asymmetric effect, capable of contracting the distances between the particles of this small body more intensely in the part closest to the most massive object that gives rise to the strong gravitational field. But how would space-time transmit to the material particles of the small body the information that it is curved or compressed in a certain region? If it were a properly physical medium in the classical sense, the information would be passed through the contact between the surface of the medium and the surface of the particle, still assuming that the particle is corporeal, endowed with surface and extension. But in the case of a purely geometric medium, a mathematically distorted vacuum, with non-corporeal structural lines, which form essentially abstract densification curves and yet energetic densification curves, how could we propose a form of physical interaction between these distorted structural lines and a set of material particles? How do particles detect the mathematical curves of space-time and the need to respect and follow them, since, in this case, we could not speak of superficial, Newtonian mechanical contact? The question becomes metaphysical, assuming contours of intimate faith, simply believing or not believing that curves of a purely mathematical nature - that cause alteration in the distribution of energy in the vacuum - can generate force to bring all the particles of a material body closer together, perhaps decreasing the length of each particle, increasing their individual and group kinetic energy, accelerating the whole body, forcing it to change its three-dimensional trajectory, as occurs with an asteroid captured by Jupiter's gravity.

Even if it is considered, mathematically, that the effect of space-time curvature causes alteration of the body's trajectory only concerning the three-dimensional component and that in four-dimensional space-time, considered as a whole, the body's trajectory describes a geodesic curve equivalent to a straight line in three-dimensional space, it seems indisputable that in the relativistic model, the curvature of space-time gives rise to a force that provides energy and acceleration directly to the particles of the material body, modifying their Newtonian behavior. We are forced to return to the question: how does this happen? What is the physical cause-and-effect mechanism that results in curved mathematical structures applying force and providing energy to particles of matter? A simple question, which not even Einstein dared to answer, probably for the simple reason that the answer is perhaps inconceivable, untenable, or lacking the explanatory simplicity that his privileged intellect so craved.

Geometry and statistics

The situation seems to get worse when we consider particles of matter as statistical entities, according to the precepts of Quantum Mechanics. We now have two mathematical entities, space-time (of a geometric nature) and quantum particles (of a statistical nature). It could be argued that while quantum particles are not observed they could occupy the entire volume of the material body, due to their ability to individually occupy infinite positions in a given temporal coordinate, according to Feynman's interpretation of Quantum Mechanics. In this case, there would be no interstitial space left that could be occupied by the space-time tissue, as a single particle could occupy the entire volume of the body, due to the concept of multi-location resulting from the principle of quantum uncertainty, in Feynman's interpretation. If a single particle could occupy the entire volume of the body, what about all of them? It is nonetheless a philosophically interesting hypothesis. All space-time would be pushed out of the material body. Even so, this path maintains or increases the enormous difficulties to formulate a satisfactory explanation for the physical mechanism of microscopic and macroscopic interaction between matter and space-time. In this descriptive scenario, the superimposition of the probability waves of all microscopic particles of a body, within the dimensional limits of that body, results in this body somehow behaving in a Newtonian way, as postulated in the quantum model itself. What turns out, curiously, is that we are Newtonian bodies only for a statistical reason. From the perspective of Quantum Mechanics, it seems that we are nothing more than statistical human beings who evolved thanks to the extinction of statistical dinosaurs, caused by the fall of a giant statistical asteroid, which had its trajectory determined by a geometric space-time, curved by our statistical planet. Nothing needs to be explained in Newtonian terms, because nothing is essentially Newtonian, since all physics is genuinely mathematical. In this light, the Newtonian forces involved, and their concrete effects, are generated by symbolic representations of reality constructed by conscious human beings composed of statistical matter, immersed in a geometric substrate. Physical reality boils down to geometric and statistical components identified or created in our conscious minds, which in turn also result, in some way, from the interaction between these mathematical components. Such components, the geometric and the statistical, were assumed in theoretical models built historically without a necessary integration between them. In an attempt at integration, it was postulated that this space-time would have smooth curvatures at the macroscopic level and frantic agitations at the quantum scales. Despite this insufficient integration between the two models, we have reached a stage in our scientific evolution in which the worst sacrilege would be to state that the universe is not an ontologically mathematical being. There is no doubt that, in the currently most accepted scientific narrative, these two components of physics (statistical and geometric) need to be integrated and unified, although it seems almost unthinkable nowadays, due to the strength of the mathematical paradigm, convince the academic community of the need to seriously and urgently investigate the hypothesis that this paradigm and its two most appreciated components simply correspond – despite their historical merit – to misconceptions under a realistic perspective of nature, not in its representative character and predictive of reality, in a degree of approximation compatible with humanity's current technological advance, but perhaps mistaken in treating as ontological reality the mathematical description that represents the perceivable behavior of nature. I particularly defend that any representation of reality is always limited, attached to a certain level of evolution of scientific knowledge, subordinated to a certain current paradigmatic system, as proposed by Thomas S. Kuhn, with unique skill and consistency.

The value of mathematics as an instrument for representing nature

Mathematics has formidable merits. It made it possible to describe the statistical behavior of microscopic particles and even anticipate the existence of some of these minuscule entities before we were able to detect them experimentally, directly or indirectly. Managed by Einstein, mathematics allowed him not only to predict the phenomenon of the gravitational deviation of light, but also, to calculate with impressive precision the exact measure of the change in trajectory that the Sun's mass causes in the light rays emitted by distant stars. Applied by Maxwell to Faraday's ideas, allowing humanity to master electromagnetism, improving the original conception of a static field to a wave field. Newton's formulations gave us the chance to enter the era of space travel. All current technology would not be possible without the precision of mathematical language. It is an extraordinary tool for the advancement of science, there is nothing to argue about that.

Mathematics and the natural world

I think, however, that we should not intend to change the nature of this valuable instrument for representing the regularities of the natural world. The mathematics we construct is limited to our conscious ability to imagine and observe the mechanisms of nature. The evolution of this language advances according to the development of this human capacity for observation and imagination. However, nature is not bound by the limits of our consciousness and can always have aspects not yet imagined or perceived by our intellect or by the intellect of artificial intelligence.

Causality in nature

I believe that it is necessary to reinforce the old idea that nature behaves causally, eventually producing regularities, and it is because of these regularities that mathematical formulations applied to physical processes work in a way that we usually consider satisfactory. Even in a chaotic environment like the movement of molecules in a gas, there are general variables that we can determine and relate. Starting from the variation of their values between two sufficiently close moments, it is possible to predict what will happen in the following moments, based on mathematical equations, which most often result from the observation of how these variables normally behave concerning each other. Prediction capability also results from the application of pre-existing mathematical models to phenomena that are being studied and modeled. In the latter case, observation produces a series of data that can remain without identifying a precise relationship between them, until someone decides to apply an old mathematical model, developed out of pure dilettantism by some restless and obstinate genius for the study of numerical properties in a context of abstract variables. Eventually, the chosen model fits the observational data like a glove or just needs just a few adjustments, arriving at one or more equations that allow an understanding of the causal relationship between the states of things at different times.

Mathematics, causality and representation

I understand the application of mathematics to physical regularities as the identification of causalities between the variables that we perceive as important in the occurrence of these phenomena. However, as we study the same phenomena on smaller and smaller microscopic scales, it seems obvious that other variables and causalities will emerge that previously remained hidden from our perception. Mathematics is like a painting made by a realist painter, who seeks to capture and represent what he sees, in the most perfect and detailed way possible, but will always have to be content with representing what is allowed by the limits of his visual acuity, as well the physical characteristics of the paints and fibers of the canvases and brushes he may use. It will have to leave out of the picture the representative image of trillions of microorganisms, cells, molecules and atomic and subatomic particles that are certainly before his eyes. It will never be able to transpose to the canvas all the minuscule elements that give structure and define the portrayed reality. The same is true of microscopes and telescopes, however powerful they may be.

The object and its representation – serious philosophical error

It seems to me of fundamental importance for a significant leap in the scientific development of humanity to clearly establish the separation between representation and represented object. Mathematical models are in the field of representation, and cannot be confused with the object they intend to portray. I regard as a colossal historical error of a philosophical nature the widespread adoption of the idea of elevating mathematical structures to the level of physical entities. This is a confusion that we must urgently overcome, as it has become an enormous barrier these days to a deeper understanding of the workings of the universe. In his book The Grand Design, Stephen Hawking very well illustrates the reduction in the importance that the scientific community has attributed to philosophy, in favor of purely mathematical conceptions, in a cartoon that shows a theoretical physicist pointing to a blackboard with an equation and telling a couple of colleagues: "This is my philosophy".[7]

No one will disagree that mathematical models often allow new philosophical inferences and the emergence of new explanatory conceptions regarding the functioning of natural mechanisms. Quantification often paves the way for revolutionary qualitative models, as happened, for example, with Lavoisier's discovery, based on

---

[7] HAWKING, MLODINOW, 2011, p. 8

exhaustive measurements of the weight of air before and after burning oxygen, that air is composed of a mixture of gases of different elements or chemical substances, replacing the paradigm of the phlogiston theory that prevailed until then. History demonstrates that sometimes theory precedes data or physical evidence, as was the case with Alfred Wegener's Theory of Continental Drift. It can be argued that he used the format of maps of the continents as the first evidence, but it is known that he had to look for more robust evidence to support his theory and make it accepted in scientific circles.

These two knowledge evolution processes – data preceding explanation and vice versa – compete and complement each other.

The explanatory model that I will share later assures me full conviction in stating that we are experiencing a historic moment in which the current impasses in physics will only be resolved with an innovative qualitative conception of the physical world, even if this innovation means taking advantage of and adjusting some concepts already proposed and rejected, recombining and developing them into a more complete descriptive and explanatory framework. I believe that this new description can fulfill the role of allowing the emergence of new mathematical models that will corroborate its general principles, paving the way for its refinement, through the definition, detailing and selection of those developments that prove to be more consistent.

The structure of space-time

A critical argument regarding space-time, which we can develop a little further, concerns the very fact that the fabric of space-time requires a constitution, a structure, with physical properties. It is worth remembering that we have already positioned ourselves against the possibility of an essentially mathematical structure of nature. As for the structural physical aspect of space-time, considering its physical capacity to interact with matter, we must imagine a space-time with some characteristics of ordinary matter. Even if we could not completely discard the mathematical nature of these two substances – matter, with its statistical structure, and space-time, with its geometric structure –, we would be obliged to admit that at some point and in some way both mathematically structured substances assume concrete material structure.

This is obvious in the case of matter, which is visible and palpable to us, but it must also apply to space-time. Without a concrete material structure, neither ordinary matter (or condensed energy) could curve space-time, nor space-time could lead matter through the gravitational curves that matter itself imprints on it since in the latter case there would be no reason for matter to respect these curves. Due to its physical properties, and according to the Einsteinian concept of *substance*, as an independent physical reality, the space-time fabric *has to be a substance*, despite its alleged mathematical composition. It cannot be a purely abstract, geometric entity, because that way it would be something devoid of any physical capacity, except for a supposed causal interference of the equations developed to quantitatively describe the behavior of the material world on this same mundane behavior.

I will repeat the previous caveat: except for a supposed causal interference of the equations developed to quantitatively describe the behavior of the material world on this same mundane behavior. It would be to say that when we describe the world, the description becomes the world. Einstein, forgive me, but this contradicts his thinking about the evolution of science, as I will show later with the parable of the closed clock formulated by Einstein himself.

To make it clearer, we build mathematical equations to describe the behavior of the physical world and when we ask why the physical world presents such behavior, we answer that the deepest cause of this behavior is its mathematical description, which starts to have absolute ontological value. This is the current philosophical problem in physics: the adoption of explanations that are ultimately tautological in nature for fundamental physical mechanisms. The representation is equated with the essence or the cause of the essence. The world has the nature of a mathematical representation model because there is a mathematical representation model of the world.

Admitting, despite its mathematical description, that space-time must be a physical substance that can interact with matter, it would probably be the only non-granulated physical entity in the entire known universe, since it is, by definition, something continuous.

Continuity, permeability, deformation, compression and elasticity (2)

We can do another little mental exercise to try to imagine again the interaction between space-time and matter. If matter deforms the surrounding space-time, the movement of an atomic matter particle, for example, will deform space-time as the particle moves, forming a gravitational micro-field, regardless of whether this particle is part of a material body. In this interaction model, such micro-deformation would, in principle, be in the nature of a micro-compression exerted on the portion of space-time that surrounds the particle of matter, moving away the spatial fabric and opening a way for the particle to cross it. Repeating a previously formulated question, how could it be formed, then, the gravitational macro-field of the material body as a whole? We can continue the exercise by imagining that while the particles that make up the material body move through space-time, assuming they are all being observed – so as not to contradict, for now, the uncertainty principle –, the resulting micro-deformations would then be transmitted, from somehow, to the most external regions of the body, to form its gravitational macro-field. This would result in a gravitational field formed by a tightly compressed space-time fabric around the material body, while this body traverses space-time in its translational trajectory.

Some problems result from this interaction model, which we could call push-away compression. In the first place, it would be logical to assume that this compression would require an energy consumption that would cause a loss of kinetic energy of the particles of matter and of the material object itself as a whole, which could compromise its inertia.

This could be resolved by a hypothetical elasticity of the space-time continuum, which would return energy to the particles, allowing the indefinite movement of material particles through space-time. Even so, some drawbacks would remain. First, we would have to admit that the normal interactions between the atomic and subatomic particles of a material body, which usually take place at a certain distance, ignore the existence of a continuous physical structure that fills the gap that exists between them. This interaction would occur between particles of matter *despite* the physical structure that encapsulates them. If matter particles had minds, many of them would have to be concerned, on this assumption, not only with pushing space-time physical structure out of their path as they move, but also in interacting at a distance with other particles of matter, which, in turn, would also be concerned with making their way through the continuous medium in which they are immersed. Interactions arising from electric charge, for example, would occur in an environment of presumably significant elastic disturbances caused by the movement of the charge-carrying particles themselves.

On the other hand, if space-time had elastic properties, the compression could be continuously neutralized, so that space-time away from it would once again occupy the previous space left by the particle in motion, as it moves. This elasticity would eventually prevent or hinder the transmission of the push-away compression to the outside of the material body. In place of large gravitational curvature, we could have an infinity of tiny jerky ripples in the space-time fabric within and on the surface of the body.

Even if a supposed compression of space-time could be transmitted to regions external to the massive body, to form the macro deformation of the continuum, this body would hardly have the gravitational properties that we know, because assuming that it is necessary to perform physical work to compress space-time in order to allow the displacement of a particle through it, it would be expected that the compressed space-time in the region of macro deformation around the massive body would require, in theory, a greater physical work to be traversed by other bodies in this region, the which would tend to make material objects move away from each other, thus creating an anti-gravitational effect. In other words, if around every particle or massive object there is a compressed space-time, possibly two bodies close to each other would tend to move away instead of approaching, due to the greater consumption of energy hypothetically necessary to cross the region of greater compression. Antigravity would be caused by the tendency of bodies or particles to move through portions of space-time with less resistance, that is, less compressed, moving away, therefore, from other bodies or particles, contrary to the behavior observed in objects with mass.

According to the reasoning that the deformation of the space-time fabric by the presence of matter implies compression of the continuum, nothing seems to indicate, strictly speaking, that massive material bodies would tend to fall into the most compressed regions of space-time, as proposed by Einstein, rather than being compelled to deviate from these regions.

The reader could again argue that this interaction model is based on mechanical concepts already superseded by the current physical field concept. Eventually, you would consider that we would have to analyze and introduce into the discussion the concept of tensors that form in the gravitational field and the variation of space-time energy along this field, represented by Einsteinian tensors. But from the point of view of looking for an underlying physical mechanism, would that make the situation better? The gravitational field, conceived as a space-time region, has variable gravitational energy inherent to space-time itself in that region, induced by the presence of a material body. The question remains much the same: how can matter energize the fabric of space-time, formed by a mathematical structure? How vectors grouped in tensors, with direction and intensity, mathematically represented, simply appear and are incorporated into the mathematical structure of space-time by the influence of matter, creating a force that, instead of pushing other bodies away, impels them towards the center of the gravitational field, contradicting the push-away compression hypothesis that I just considered?

The bottom line we want to discuss is that General Relativity does not seem to have formulated any descriptive model, at the microscopic level, of the real physical interaction between matter and space-time, compatible with observable gravitational effects, and this consists of a gap we must try to fill to save the fabric of space-time as a physical reality, or sooner or later we will be forced to abandon the concept of space-time. It will be remembered that General Relativity superseded the mechanical model only conditionally, and the implicit condition is that the successor theory filled in its gaps, without the need to resort to the superseded theory. If Newtonian concepts prove to be important tools for testing the limits of the space-time concept, nothing should stop us from using them, even more so in the absence of any relativist tool that proves indisputable. In this case, the argument that the tool used is simply outdated does not seem valid, as this may prove to be true only for some situations and not necessarily for all.

Likewise, Quantum Mechanics does not seem to be a theory from which the *physical* mechanisms responsible for electrical interaction result. It does not emerge from this theory what positive and negative electric charges are, or why some particles have a charge of a certain kind and others have no charge at all. On the interpretation of Quantum Mechanics (emphasis added):

When a measurement is performed on a quantum system, the result provides new information about the system in question. This implies a modification of the probabilities associated with the possible results of a subsequent measurement. Since these probabilities can be deduced from knowledge of the quantum state, it can be concluded that measurement affects the quantum state in a way that depends on the result obtained. In fact, in the simplest case, the result obtained completely determines the subsequent quantum state. This change of quantum state by measurement is called quantum state reduction or collapse.

In the case of the Stern-Gerlach experiment, if the detector located on the upward deflection path fires, then, after firing, we will know that the atom has an upward spin. Consequently, the spin of this atom at the output of the apparatus will be described by the state $|+)$. Similarly, if the detector located on the downward deflection path fires, the subsequent spin state will be $|-)$. That is, the initial general pure state will be reduced to $|+)$ or $|-)$, depending on which of the detectors fires.

*In fact, we don't even need two detectors to operate the complete quantum state reduction. If we place a detector, for example on the deflection path downwards, then if the passage of an atom through the device does not cause the detector to trigger, we can conclude that the atom has not been deflected downwards, and therefore has been deflected upwards. In possession of this information, we can say that the atom has spin up, and we will describe it by the state $|+)$. **In this case, the state of the atom will be reduced without any apparent interaction between the atom and the detector!***

After the above discussion, it is clear that the reduction of the quantum state must be conceived as a logical process linked to the acquisition of new *information, and not as a strictly physical process. **For this reason, the search for an interpretation of quantum states as representing an underlying objective reality - a philosophical attitude known as realism - is hardly reconcilable with Quantum Mechanics in its usual formulation.**[8]*

In short, what I wanted to emphasize in the last paragraphs is that there seems to be no Einsteinian or quantum model that satisfactorily and realistically explains the nature of the microscopic interactions between matter and the space-time fabric, in such a way that these result in gravitational or electromagnetic effects. The relativist way out is to offer a fundamentally geometric explanation, that acts in the region outside massive bodies, attributing physical properties to geometry. On the other hand, the attempt to explain the nature of the interaction between matter and a continuous space-time fabric runs into serious difficulties, such as the presumed need for this fabric to present certain elastic and dynamic properties that make it difficult to reconcile the permeability of the medium and its continuity, between the continuity of the medium and the inertia of particles and material bodies, and between the continuity of the medium and the formation of its macroscopic gravitational curvature. It would be necessary to construct a model that would require very sophisticated – eventually inconceivable – permeability dynamics for a physical medium (space-time), of which most basic microscopic properties have not yet been sufficiently modeled, despite the efforts of Quantum Mechanics scholars, which seems to fully justify not discarding some consecrated classical mechanical concepts to highlight the huge gaps in the microscopic interaction mechanisms between matter and the space-time fabric. Without dispensing with the advances of Special Relativity and General Relativity, especially the mathematical results of the gravitational effect and especially the relativity of time and distance measurements, we believe that a continuous physical structure, privileged over all others, is disposable, the only one to break such an eloquent rule of the natural world, that physical structures are not continuous, but made up of particles.

---

[8] http://www.if.ufrgs.br/~betz/quantum/SGtexto.htm

Einstein's view on the juxtaposition of fields

We know that Einstein concluded that the vacuum contains residual energy, a kind of latent heat, and we know that he helped to demonstrate that some forms of energy such as heat and light are transmitted by tiny particles, the quanta. Following the line of reasoning that we are defending, we can ask if this vacuum energy is inherent to the spatial fabric or something that inhabits it without composing its structure. Would it be possible to maintain that this latent energy has a geometric origin? Or should we apply Ockham's razor and assume that it is simpler and more likely that this latent energy is manifested and conserved through the movement of particles? The mere possibility of the question is yet another clue that the "space" that interacts with matter can be "constituted" by particles (actually a dynamic ether that occupies an inert space).

On the other hand, we observe that Einstein only admitted two distinct physical realities: matter (energy) and the field (of force, of the gravitational and electromagnetic types, specifically), although he expressed his absolute conviction that both would one day be unified by the evolution of scientific knowledge. Where would the space-time fabric be located, then, in the context of physical realities, that is, of substances? Would it be more for matter or field? Or for both? Or none? By imagining the unification of matter and field, Einstein even speculated, with Leopold Infield, that <u>matter</u> could be "the regions of <u>space</u> where the <u>field</u> is extremely strong" [9] – Quantum Field Theory (quantum gravity) seems to have explored this path. In such a hypothesis there would be a triple unification: matter, field and space. But before talking about a unified theory, we need to be able to clearly answer what would be, physically, space. General Relativity offers a permeable continuum, a geometric substance devoid of particles, a dormant, latent field with a standby energy of its own, waiting to be locally converted somehow into a source of electromagnetic or gravitational force. In my view, all this is excessively complex, contradictory and unsatisfactory. There is a simpler explanation for how the universe works!

Conjectures aside, it seems impossible to me to avoid a feeling of veneration to the genius and the soul of a human specimen who not only

---

[9] EINSTEIN, INFELD, 2008, p. 202

built incredible shortcuts for the advancement of knowledge, but was also able to express absolute honesty of purpose in exposing with moving precision and clarity the dilemma that his own logic imposed on him, clearly with the sole intent of paving the way for those who would follow him and inspire them to seek scientific truth.

When referring to the triumph of the association between gravitation and the geometric structure of space circumscribed to the field of gravitation, and alluding to the fact that the results of experimentation prove the mathematical predictions of his gravitational theory, these were his exact words:

> *It is not a matter of discussing here the question of verifying the theory by experience, but of clarifying immediately why the theory cannot be satisfied with this result. Gravitation was reintroduced into the structure of space. It is the first point, but outside this gravitational field, there is the electromagnetic field. It will first be necessary to theoretically consider this last field as a reality independent of gravitation. In the conditional equation for the field, I was constrained to introduce supplementary terms to explain the existence of this electromagnetic field. But my theoretician's spirit absolutely cannot support the hypothesis of two structures of space, independent of each other, one in metric gravitation, the other in electromagnetic. My conviction stands: the two kinds of fields must actually correspond to a unitary structure of space.*[10]

Before proceeding, we need to pause, just to look at the depth of the motivation for this appeal. It is as if one of the greatest names in science of all time, and certainly the greatest in physics in the last three centuries, were extending a hand to humanity and at the same time asking for help to decipher it, not the result of an enigma, which already foresaw, but the way to arrive at it.

By stating his deep conviction, he seems to want to emphasize the importance of committing ourselves to discover the truth according to validation criteria that satisfy the restlessness of our spirits, having

---

[10] EINSTEIN, 1981, pp. 175-176

offered us, however, the demarcation of a clear limit to explore, a reference, his best guess, as a legacy.

We can also see in this statement by Einstein an alert to the fact that the experimental proof of the mathematical predictions of a theory does not guarantee the correctness of the physical premise adopted as an explanation of the results obtained.

Einstein's view on the evolution of physics

Einstein seems to be an inexhaustible source of reflections on the evolution of physics as a creation of the human mind. In this regard, in partnership with Infield, he taught:

> *Physical concepts are free creations of the human mind, not being, as more as it may seem, uniquely determined by the external world. In our effort to understand reality, we are something like a man trying to understand the mechanism of a closed clock. He sees the dial and hands moving, he even hears their ticking, but he has no way of opening the case. If he is ingenious, he may form some image of a mechanism that could be responsible for all the things he observes, but he can never be quite sure that his image is the only one capable of explaining his observations. He will never be able to compare this image with the real mechanism, and he cannot even imagine the possibility or the real meaning of such a comparison. But he certainly believes that, as his knowledge increases, his picture of reality will become more and more simple and will explain an ever-increasing range of sensory impressions. He may also believe in the existence of the ideal limit of knowledge and that the human mind approaches it. You can call this ideal limit objective truth.[11]*
>
> *(...)*
>
> *In our great mystery story, there are no problems completely solved and solved forever. After 300 years we*

---

[11] EINSTEIN, INFELD, 2008, p. 36

had to go back to the initial problem of motion, to review the investigation procedure, to find new clues that had been overlooked, thus arriving at a new picture of the universe.[12]

How can we not admire, from the bottom of our souls, a mind like this? I came into direct contact with Einstein's words looking for something insurmountable that would make the foundations of the Algorithmic Matter Theory collapse in its rudimentary beginnings, but what I found was an overview of the current development of physics, which made me conclude that the introduction of Algorithmic Theory is not only viable, plausible, but inexorable.

---

[12] EINSTEIN, INFELD, 2008, p. 39

CHAPTER II
THE ALGORITHMIC MATTER

Algorithmic Matter Theory

The Algorithmic Matter Theory proposes a universe populated by particles that form a medium, properties of which are capable of creating and sustaining matter, producing the effects of gravity and inertia, causing electromagnetic manifestations, explaining the zoo of subatomic particles, as well as the supposed dual (wave-particle) nature of both light and atomic matter components, the supposed quantum entanglement, the distortion of time and distance measurements, and virtually every other physical phenomenon from dark matter and reciprocal conversion between matter and energy to the formation and evolution of biological processes and of consciousness itself, projecting what may be the basis for a more comprehensive understanding of the functioning of the universe. If it is possible to explain a range of phenomena of such magnitude from the properties of an ethereal medium, without it being necessary to start with a single mathematical symbol, it is very likely that this medium exists and has such properties, although the inventive spirit of human beings, combined with the advancement of scientific knowledge, can always find, as in the example of Einstein's closed clock, other explanations for the same phenomena.

New possible assumptions

Let us then go on to present other aspects of the theory, which involve the specific properties of the medium that constitutes its basic premise. The extreme simplicity of the fundamental postulate of Algorithmic Theory contrasts with the breadth and complexity of the questions it purports to help elucidate. Its basic postulate is so simple that its mere revelation perhaps makes clear the moment when physics, in making the Einsteinian leap, left something important behind. Something that could be the natural evolution of the Cartesian ether, Newton's kinetic model, Huygens' wave model, and Faraday and Maxwell's electromagnetic scheme, providing the conception of a theoretical physic's foundation to rival Einstein's revolutionary and much welcomed relativistic model. But physics evolves historically and in leaps. Without General Relativity, we would not have conceived of algorithmic matter nearly a century later. We can say that General Relativity is the mother of AMT for two reasons. First, because AMT arose out of a deep logical uneasiness caused by the relativistic idea about the nature of space-time. Second, because AMT not only accepts and presupposes the relativity of the measures of time and distance but in a reverse way, from the new model of reality created, makes such relativity necessary and evident. The new model deepens this relativity to the point where the very premise of the constancy of the speed of light, fundamental in the Einsteinian model, is also relativized, in the sense that such constancy can be conceived as necessary only from the local point of view, opening up the possibility that, as it traverses the universe, light has its speed successively modified by the different gravitational conditions it encounters in its trajectory when considering the parameters of time and distance associated with the local conditions of a very distant observer.

Particularly, AMT seeks to explain the "curvature of space" as an effect of the dynamics between particles of known matter and energy and particles of the algorithmic ether, which can be conceived as dark matter ether. So far, we have left out dark energy. The reason is that we have not yet dedicated ourselves to elaborating a conceptual formulation for it, although we believe that its manifestation may result from the decrease in the density of first-order dark matter – a term that we will explain later –, in the emptiest regions of the universe, as a consequence of the hypothetical expansion of the known universe and its algorithmic dynamics.

The beginnings of the universe

We do not intend to discuss, for the time being, the origin, or the exact reason for the admitted current expansion of the universe, because, in the context of AMT, I also do not yet have a well-developed proposition for this beginning, other than to attribute an algorithmic nature to the eventual event that initiated the current hypothetical expansionist phase. At the moment, I consider this question secondary, as much as the question of dark energy, to the more immediate objective of the theory, which is to explain the functioning of the universe, in themes such as gravitation and the nature of microscopic particles. To begin explaining the algorithmic model we are proposing, we start from the assumption that at the beginning of the apparent accelerated cosmic expansion underway, and before it, there were primordial particles, on size scales much smaller than those involving the particles of matter with which we are used to dealing. The supposedly expanding portion of matter and space that we call the universe, an expansion currently recognized by most of the academy, could be very hot, and the primordial particles moved freely with frantic speeds, in a way that we would consider random, chaotic, with all type of interaction between them. It is admitted that the phase had a lot of energy, particles were created, annihilated, absorbed, and transformed. During the expansion, the universe began to cool down and little by little the interactions between the primordial particles began to show some flow patterns, what we might call *natural motion processing algorithms*. In a Darwinian universe, various kinds of flow algorithms came and went, destroyed by more powerful and stable algorithms, produced and eventually sustained by the dynamics of changing cosmic conditions that followed. Some of these algorithms gave rise to new particles, with a dimensional scale superior to that of the primordial particles. The universe was full of no punctiform particles, structured by continuous movement, in flux, perhaps some in the form of vortices, formed by much smaller particles compared to the size of each vortex or other type of motion algorithm.

Successively, particles of a certain dimensional scale began to be engulfed by a swirling flow, or another flowing pattern of cellular architecture, and this flow constituted another particle, of a higher dimensional scale. Particles of similar size and energy on a given dimensional scale, but with incompatible or antagonistic flow algorithms, may have arisen naturally and collided with each other, in the phenomenon currently known as the mutual annihilation of matter and antimatter, until the stabilization derived from the quantitative predominance of one strain over the other. In each dimensional scale, and in each temporal phase of cosmic evolution, surviving algorithms, with a high degree of stability and occurrence, evolved into the most common species of particles that exist today. At different scales evolution need not have followed the same pace and pattern as physical conditions evolved, because changing conditions may have been determinant for some scales and not for others, concerning stable and dominant particle forms.

If the fundamental particles of each dimensional scale are formed by the rotatory motion (or another pattern of cellular flow) of smaller particles, in a chain of indefinite successive scales, it would be fair to ask what the primordial particles were formed of. This doesn't really matter right now, and the AMT doesn't offer any guarantees that we'll ever get the answer, but two logical possibilities open up more vigorously. In the first hypothesis, the primordial particles themselves were also formed by the algorithmic movement of particles of even smaller scales, in an infinite progression, and in this case, the primordial particles could be seen as those of maximum size in the initial conditions of what is considered the current stage of expansion of the universe. In the second case, on some very small scale, almost unimaginably smaller than the scale of primordial particles, there could be a limit, and particles of that scale would have the minimum size that the whole universe could support, under the conditions corresponding to a certain phase of its evolution. In the latter case, the truly primordial particles, of the smallest possible scale, would not be formed by the sustained movement of smaller particles, and we would have to admit the existence of continuous fundamental particles. This last possibility would be subject to the same criticism that we present regarding the continuity of the space-time fabric and seems to us very remote and improbable, ultimately contradicting the algorithmic principle that we are about to present, which is a recursive principle, in a sense adapted from the term, as this recursion tends to be imperfect. Because it is recursive, the algorithmic principle dispenses, by definition, the limitation of size and logically consists of the hypothesis of infinite scalar progression, towards both smaller and larger scales than the perceptible objects in the currently observable universe. In this way, we must admit infinity and choose to consider the primordial particles referred to as belonging to a scale of maximum magnitude in the initial conditions prevailing in the cosmic portion currently considered as our expanding universe.

The human scale

But why? Why should primordial particles be considered to be of maximum size? Because we are characterizing the evolution of objects at a certain size scale, our scale. This scale ranges from the size of known subatomic particles to large clusters of galaxies, and even to what is currently considered the entire known universe. It is the scale of objects directly or indirectly observed by humans. We can call this scale the human scale or the anthropocentric scale. It is assumed that on this vast scale of the universe, the largest objects evolved from the grouping and combining of the simplest particles, say, quarks and electrons. But, as we shall see, according to the algorithmic principle, even quarks and electrons are constituted by the continuous motion of an infinity of much smaller particles. If in the early days of the universe at our scale there were only particles much smaller than quarks and electrons, those were the largest particles in the early phase of the universe at our scale and only later, by their combination and clustering, were larger particles and objects formed, but then we are no longer talking about the early, primordial phase.

Dimensional scales and fundamental particles

For AMT, it makes no sense to talk about punctiform physical particles. Even the most fundamental particles of any dimensional scale are endowed with some extension. I consider a geometric point a mathematical idealization, devoid of any physical properties. I don't believe that something physical can be dimensionless, one-dimensional, or even two-dimensional, on its most intimate scales.

We are introducing a new philosophical conception. The space limited to our universe and, depending on the semantic scope of the term, beyond it, is occupied, filled, by a *medium* formed by particles of smaller and smaller dimensional scales. The matter of our scale also forms a medium that fills the space, together with its interstices, and can serve as raw material for algorithmic particles of the scale immediately superior to ours, and so on infinitely. We postulate that at each scale, the particles are constituted by the algorithmic movement of particles of smaller scales. These algorithms consist of flow patterns subject to a set of possible changes according to the conditions of interaction with the medium. Philosophically, we do not accept the unattractive hypothesis of the existence of a minimal and continuous primordial particle, even if it were unimaginably far from the scale on which we live. The nature of all particles at all infinite scales is algorithmic, in the sense that its existence depends on the translative motion in a continuous cyclic flow of sustaining particles present in the cosmic medium, which can be drawn into and expelled from the flow like air molecules in a tornado. At a certain dimensional scale, different species of particles, of different sizes, defined by different motion processing algorithms (cellular flow patterns) can coexist. These algorithms (particles) can interact with each other, combining to form a more complex algorithm, which keeps the particles (flow units) that compose it cohesive, modifying them or not. Thus, particles can be formed by a unitary structure of the cyclic flow, like some kind of vortex or some kind of as-yet unimagined flow cell, or they can constitute a combination of unitary structures, forming a more sophisticated composite structure. The simplest structures that make up a composite structure particle may or may not be able to break free, under certain conditions, and survive autonomously and stably in the algorithmic ether. In theory, nothing prevents elementary particles of a dimensional scale from being of the composite type, made up of simpler algorithmic units. At each scale, the smallest particles, whether unitary or composite, can be considered elementary, fundamental particles. There may be doubts about the dimensional scale in which a given particle should be framed, and it may eventually be more appropriately classified in an intermediate scale between two sharper adjacent scales. These, among many others, are just a few hypotheses that arise from the algorithmic principle.

The algorithmic principle

In case the previous topic was not clear enough about which algorithmic principle we are introducing, it is necessary to conceptualize and rescue the history.

The algorithmic principle is the name that I am giving to the old general idea that matters resulting from the cyclic motion of ether particles.

For Descartes, the ether has a maximum particle size, defined by God at the time of the creation of ether. God set the particles in motion and from that moment they began to collide, wear out and break apart and some arising forms and flow patterns began to create, transform and sustain the known forms of matter. In this Cartesian model, no vacuum is possible, as the smaller particles fill all the interstitial space formed by the contact between the larger particles and so on. There is no minimum particle size, but some special shapes that give origin and meaning to the material world. At this point, I feel obliged to insert, in full, the most complete and summarized description that I have found about the different species of mechanical ether conceived throughout the history of physics.

Mechanical explanations of gravitation

Mechanical explanations of gravitation (or kinetic theories of gravitation) are attempts to explain the action of <u>gravity</u> through basic mechanical processes, such as forces of <u>pressure</u> caused by <u>impulses</u>, without using any <u>action from a distance</u>. These theories were developed from the 16th to the 19th century in connection with the <u>ether</u>. However, these models are no longer considered viable theories in the mainstream scientific community and the <u>General Relativity</u> is now the standard model for describing gravitation without using actions at a distance. The modern hypotheses of "<u>quantum gravity</u>" also try to describe gravity by more fundamental processes such as particle fields, but they are not based on classical mechanics.

Shield

See the main article: <u>Le Sage's gravitational theory</u>

This theory is probably[1]the best-known mechanical explanation and was first developed by <u>Nicolas Fatio de Duillier</u>in (1690) and reinvented, among others, by <u>Georges-Louis Le Sage</u> (1748), <u>Lord Kelvin</u> (1872), and <u>Hendrik Lorentz</u> (1900), and criticized by <u>James Clerk Maxwell</u> (1875) and <u>Henri Poincaré</u> (1908).

The theory posits that the <u>force</u> of gravity is the result of small <u>particles</u> or <u>waves</u> moving at high speed in all directions, all over the <u>universe</u>. It is assumed that the intensity of particle flow is the same in all directions;

therefore, an isolated object A is hit equally from all sides, resulting in only one pressure directed inward, but no net directional force. With a second object B present, however, a fraction of the particles that would have hit A from the direction of B are intercepted, so B acts as a shield, so to speak - that is, in the direction of B, A will be hit by fewer particles than in the opposite direction. Likewise, B will be hit by fewer particles from the direction of A than from the opposite direction. It can be said that A and B are "shadowing" each other, and the two bodies are pushed against each other by the resulting imbalance of forces.

This shadow obeys the inverse square law, because the imbalance of momentum flux over an entire spherical surface surrounding the object is independent of the size of the surrounding sphere, while the area of the sphere increases proportionally to the square of the radius. To satisfy the need for mass proportionality, the theory postulates that: a) the basic elements of matter are very small so that bulk matter consists mostly of empty space, and b) that particles are so small that only a small fraction of them would be intercepted by bulk matter. The result is that the "shadow" of each body is proportional to the surface of each element of matter.

Criticism: This theory was rejected mainly for reasons of thermodynamics because a shadow only appears in this model if the particles or waves are at least partially absorbed, which should lead to enormous heating of the bodies. In addition, drag, i.e., the resistance of particle flows in the direction of motion, is also a big problem. This problem can be solved by assuming speeds larger than light, but this solution greatly increases the thermal problems and contradicts Special Relativity. [2][3]

Vortex

Because of his philosophical beliefs, René Descartes proposed, in 1644, that no space emptiness can exist, and this space must consequently be filled with matter. The parts of this matter tend to move in straight paths, but because they are close together, they cannot move freely, which, according to Descartes, implies that all motion is circular, so the ether is filled with vortices. Descartes also distinguishes between different shapes and sizes of matter, in which bulk matter resists circular motion more strongly than ethereal matter. Because of centrifugal force, the matter tends to the outer edges of the vortex, which causes condensation there. Rough matter cannot follow this motion because of its greater inertia − therefore, due to the pressure of condensed external matter, these parts will be pushed towards the center of the vortex. According to Descartes, this internal pressure is nothing more than gravity. He compared this mechanism to the fact that if a rotating vessel filled with liquid is stopped, the liquid will continue to rotate. Now, if we drop small pieces of lightweight material (e.g. wood) into the vase, they will move towards the middle of the vase.[4][5][6]

Following Descartes' basic premises, Christiaan Huygens between 1669 and 1690 designed a much more accurate model of the vortex. This model was the first theory of gravitation that was worked out mathematically. He assumed that the ether particles were moving in all directions but were thrown back at the outer edges of the vortex, and this caused (as in Descartes' case) a greater concentration of fine matter at the outer edges. So also, in his model, the fine matter presses the coarse matter in the center of the vortex. Huygens also found that the centrifugal force is equal to the force acting towards the center of the vortex (centripetal force). He also postulated that bodies must consist mostly of empty space so that ether can easily penetrate bodies, which is necessary for mass proportionality. He further concluded that the ether moves much faster than falling bodies. At this time, Newton developed his theory of gravitation, which is based on attraction, and although Huygens agreed with the mathematical formalism, he said that the model was insufficient due to the lack of a mechanical explanation of the law of force. Newton's discovery that gravity obeys the <u>inverse square law</u> surprised Huygens, and he tried to take this into account by assuming that the speed of the ether is lower at greater distances.[6][7][8]

Criticism: Newton opposed the theory because the <u>drag</u> should lead to noticeable deviations from the orbits that were not observed.[9]Another problem was that the <u>moons</u> often move in different directions, against the direction of movement of the vortex. Furthermore, Huygens' explanation of the inverse square law is <u>circular</u> because that means the ether obeys <u>Kepler's third law</u>. But a theory of gravitation must explain these laws and must not presuppose them.[6]

Several British physicists developed the <u>vortex theory of the atom</u> at the end of the 19th century. However, the physical <u>William Thomson, 1st Baron Kelvin</u>, developed a quite different approach. While Descartes had delineated three kinds of matter - each linked respectively to the emission, transmission and reflection of light - Thomson developed a theory based on a unitary continuum.[10]

Chains

In a letter of 1675 to <u>Henry Oldenburg</u>, and later to <u>Robert Boyle</u>, Newton wrote the following: [Gravity is the result of] "a condensation causing a flow of ether with a corresponding thinning of the density of the ether associated with an increase in the velocity of the flow." He also stated that this process was consistent with all his other works and with Kepler's Laws of Motion.[11]Newton's idea of a pressure drop associated with an increase in flow velocity was formalized mathematically as <u>Bernoulli's principle,</u> published in Daniel Bernoulli's book Hydrodynamic in 1738.

However, although he later proposed a second explanation (see the section below), Newton's comments on this question remained ambiguous. In the third letter to Bentley, in 1692, he wrote:[12]

It is inconceivable that inanimate brute matter should, without the mediation of anything other than material, operate and affect other matter, without mutual contact, as it must do if gravitation in Epicurus's sense is essential and inherent in it. And that's one of the reasons I wish you didn't attribute "innate gravity" to me. That gravity should be innate, inherent and essential to matter, so that one body can act on another at a distance, through a vacuum, without the mediation of anything, through which its action and strength can be transmitted from one to another, to me, is such an absurdity that I believe that no man who has in philosophical matters a competent faculty of thought can fall into it. Gravity should be caused by an agent acting constantly according to certain laws; but whether this agent is material or immaterial, I left to the consideration of my readers.

On the other hand, Newton is also well known for the phrase Hypotheses non fingo, written in 1713:

I still haven't been able to discover the reason for these properties of gravity from phenomena, and I don't pretend hypotheses. For what is not deduced from phenomena must be called a hypothesis; and hypotheses, whether metaphysical or physical or based on occult or mechanical qualities have no place in experimental philosophy. In this philosophy, particular propositions are inferred from phenomena and then made general by induction.

And according to the testimony of some of his friends, like Nicolas Fatio de Duillier or David Gregory, Newton thought that gravitation was based directly on divine influence.[8]

Similar to Newton, but mathematically in more detail, Bernhard Riemann assumed in 1853 that gravitational ether is an incompressible fluid and normal matter represents drains of flow in this ether. Therefore, if the ether is destroyed or absorbed in proportion to the masses within the bodies, a current arises and carries all the surrounding bodies towards the central mass. Riemann speculated that the absorbed ether is transferred to another world or dimension.[13]

Another attempt to solve the energy problem was made by <u>Ivan Osipovich Yarkovsky</u> in 1888. Based on his ether current model, similar to Riemann's one, he argued that absorbed ether could be converted into new matter, leading to a massive increase in celestial bodies.[14]

Criticism: As in the case of Le Sage's theory, the unexplained disappearance of energy violates the <u>energy conservation law</u>. There must also be some obstacle, and no process leading to the creation of matter is known.

Static pressure

Newton updated the second edition of Optics (1717) with another theory of gravity of the mechanical ether. Contrary to his first explanation (1675 - see Currents), he proposed a stationary ether that gets thinner and thinner near celestial bodies. In the analogy of <u>support</u>, a force arises that pushes all bodies towards the central mass. He minimized drag by stating an extremely low density of the gravitational ether.

Like Newton, <u>Leonhard Euler</u> assumed in 1760 that the gravitational ether loses density according to the inverse square law. Like others, Euler also assumed that to maintain the proportionality of mass, matter consists mainly of empty space.[15]

Criticism: Newton and Euler gave no reason why the density of this static ether should change. Furthermore, <u>James Clerk Maxwell</u> pointed out that, in this "hydrostatic" model, *"the state of stress ... which we must assume exists in the invisible medium is 3000 times greater than that which the strongest steel could withstand"*.[16]

Waves

<u>Robert Hooke</u> speculated in 1671 that gravitation is the result of all bodies emitting waves in all directions through the ether. Other bodies, which interact with these waves, move towards the source of the waves. Hooke saw an analogy to the fact that small objects on a disturbed surface of water move towards the center of the disturbance.[17]

A similar theory was mathematically worked out by <u>James Challis</u> from 1859 to 1876. He calculated that the case of attraction occurs if the wavelength is large compared to the distance between the gravitational bodies. If the wavelength is small, the bodies repel each other. By a combination of these effects, he also tried to explain all other forces.[18]

Criticism: Maxwell objected that this theory requires a constant production of waves, which must be accompanied by an infinite consumption of energy.[19]Challis himself admitted that he had not reached a definitive result due to the complexity of the processes.[17]

Pulsation

<u>Lord Kelvin</u> (1871) and <u>Carl Anton Bjerknes</u> (1871) assumed that all bodies pulsated in the ether. This was in analogy to the fact that if the pulsation of two spheres in a fluid were in phase, they would attract each other; and if the pulsation of two spheres is not in phase, they will repel each other. This mechanism was also used to explain the nature of <u>electric charges</u>. Among others, this hypothesis was also examined by <u>George Gabriel Stokes</u> and <u>Woldemar Voigt</u>.[20]

Criticism: To explain universal gravitation, we are forced to assume that all pulsations in the universe are in phase - which seems very implausible. Furthermore, the ether must be incompressible to ensure that the attraction also occurs at greater distances.[20]And Maxwell argued that this process must be accompanied by a permanent new production and destruction of ether.[16]

Other historical speculations

In 1690, Pierre Varignon assumed that all bodies are exposed to impulses by ether particles from all directions and that there is some kind of limitation on a certain distance from the earth's surface which cannot be passed by particles. He assumed that if a body was closer to Earth than the limiting horizon, the body would experience a greater thrust from above than from below, causing it to fall toward Earth.[21]

In 1748, Mikhail Lomonosov assumed that the effect of the ether is proportional to the complete surface of the elementary components of which matter consists (similar to Huygens and Fatio before him). He also assumed an enormous penetrability of bodies. However, no clear description was given by him of how exactly the ether interacts with matter so the law of gravitation arises.[22]

In 1821, John Herapath tried to apply his co-developed model of the kinetic theory of gases in gravitation. He assumed that the ether is heated by the bodies and loses density so that other bodies are pushed towards these regions of lower density.[23]However, it was shown by Taylor that the reduced density due to thermal expansion is compensated by the increase in the velocity of the heated particles; therefore, no attraction arises.[17]

Recent Theorizing

These mechanical explanations for gravity never gained wide acceptance, although such ideas continued to be studied occasionally by physicists until the early 20th century, bywhich time it was generally considered that such explanations were conclusively discredited. However, some researchers outside the scientific mainstream are still trying to discover some consequences of these theories.

Le Sage's theory was studied by Radzievskii and Kagalnikova (1960),[24] Shneiderov (196),[25] Buonomano e Engels (1976),[26] Adamut (1982),[27] Jaakkola (1996),[28] Tom Van Flandern (1999),[29] and Edwards (2007). A variety of Le Sage's models and related topics are discussed in Edwards et al.[31]

Gravity due to static pressure has recently been studied by Arminjon.[32]

References

1.   ↑ Taylor (1876), Peck (1903), secondary sources

2.   ↑ Poincaré (1908), Secondary sources

3.   ↑ Maxwell (1875, Atom), Secondary sources

4.   ↑ Descartes, R. (1824–1826), Cousin, V., ed., «Les principes de la philosophie (1644)», Paris: F.-G. Levrault, Oeuvres de Descartes, 3

5.   ↑ Descartes, 1644; Zehe, 1980, pp. 65–70; Van Lunteren, p. 47

6.   ↑ Ir para:a b c Zehe (1980), Secondary sources

7.   ↑ Huygens, C. (1944), Société Hollaise des Sciences, ed., «Discours de la Cause de la Pesanteur (1690)», Den Haag, Oeuvres Complètes de Christiaan Huygens, 21: 443–488

8.   ↑ Ir para:a b Van Lunteren (2002), Secondary sources

9.   ↑ Newton, I. (1846), Newton's Principia : the mathematical principles of natural philosophy (1687), New York: Daniel Adee

10.   ↑*«The Vortex Atom: A Victorian Theory of Everything». Centaurus (in English). 44: 32–*

114.*ISSN0008-8994.   It   hurts:10.1034/j.1600-0498.2002.440102.x*

11.   ↑I. Newton, letters quoted in detail in The Metaphysical Foundations of Modern Physical Science by Edwin Arthur Burtt, Double day Anchor Books.

12.   ↑http://www.newtonproject.ox.ac.uk/view/texts/normalized/THEM00258Newton, 1692, 4th letter to Bentley

13.   *↑Riemann, b. (1876), Dedekind, R.; Weber, W., eds., «Neue mathematische Prinzipien der Naturphilosophie», Leipzig, Bernhard Riemanns Werke und Gesammelter Nachlass: 528–538*

14.   *↑Yarkovsky, IO(1888), Hypothese cinetique de la Gravitation universelle et connexion avec la formation des elements chimiques, Moscow*

15.   *↑Euler, L.(1776), Briefe an eine deutsche Prinzessin, Nr. 50, 30. August 1760, Leipzig, pp. 173–176*

16.   ↑*Go to:TheB*Maxwell (1875, Attraction), Secondary sources

17.   ↑*Go to:TheBc*Taylor (1876), Secondary sources

18.   *↑Challis, J.(1869), Notes of the Principles of Pure and Applied Calculation, Cambridge*

19.   ↑Maxwell (1875), Secondary sources

20.   ↑*Go to:TheB*Zenneck (1903), Secondary sources

21.   *↑Varignon, P.(1690),New conjectures on the Pesanteur, Paris*

22.   *↑Lomonosow, M.(1970), Henry M. Leicester, ed.,«On the Relation of the Amount of Material and Weight (1758)», Cambridge: Harvard University Press, Mikhail Vasil'evich Lomonosov on the Corpuscular Theory: 224–233*

23.   *↑Herapath, J.(1821),«On the Causes, Laws and Phenomena of Heat, Gases, Gravitation», Paris,Annals of Philosophy, 9: 273–293*

24. ↑Radzievskii, VV & Kagalnikova, II (1960), «The nature of gravitation», Vsesoyuz. Astronom.-Geodezich. obsch. Byull., 26 (33): 3–14 A rough English translation appeared in a US government technical report: FTD TT64 323; TT 64 11801 (1964), Foreign Tech. Div., Air Force Systems Command, Wright-Patterson AFB, Ohio (reprinted in Pushing Gravity)

25. ↑Shneiderov, AJ (1961), «On the internal temperature of the earth», Bollettino di Theoretical Geophysics ed Applicata, 3: 137–159

26. ↑Buonomano, V. & Engel, E. (1976), «Some speculations on a causal unification of relativity, gravitation, and quantum mechanics», Int. J. Theor. Phys., 15(3): 231–246,*Bibcode:1976IJTP...15...231B,It hurts:10.1007/BF01807095*

27. ↑Adamut, IA (1982), «The screen effect of the earth in the TETG. Theory of a screening experiment of a sample body at the equator using the earth as a screen», Nuovo Cimento C, 5 (2): 189–208,*Bibcode:1982NCimC...5...189A,It hurts:10.1007/BF02509010*

28. ↑Jaakkola, T. (1996),*«Action-at-a-distance and local action in gravitation: discussion and possible solution of the dilemma»(PDF), Apeiron, 3 (3–4): 61–75*

29. *↑Van Flandern, T.(1999), Dark Matter, Missing Planets and New Comets 2nd ed. , Berkeley: North Atlantic Books, pp. Chapters 2–4*

30. ↑Edwards, M.R. (2007),*«Photon-Graviton Recycling as Cause of Gravitation»(PDF), Apeiron, 14(3): 214–233*

31. ↑Edwards, MR, ed. (2002), Pushing Gravity: New Perspectives on Le Sage's Theory of Gravitation, Montreal: C. Roy Keys Inc.

32. ↑Mayeul Arminjon (November 11, 2004), «Gravity as Archimedes´ Thrust and a Bifurcation

in that Theory», Foundations of Physics, 34 (11): 1703–1724,*Bibcode:2004FoPh...34.1703A,arXiv:physics/0404 103,It hurts:10.1007/s10701-004-1312-3*

Bibliography

• *Aiton, EJ (1969), "Newton's Aether-Stream Hypothesis and the Inverse Square Law of Gravitation", Annals of Science, 25 (3): 255–260,It hurts:10.1080/00033796900200151*

• *Carrington, Hereward (1913), Sugden, Sherwood J.B., ed.,«Earlier Theories of Gravity»(PDF), The Monist, 23(3): 445–458,It hurts:10.5840/monist19132332*

• *Drude, Paul (1897),«Ueber Fernewirkungen», Annalen der Physik, 298 (12): I–XLIX,Bibcode:1897AnP...298D...1D,It hurts:10.1002/andp.18972981220*

• *Hall, Thomas Proctor(1895), «Physical Theories of Gravitation», Proceedings of the Iowa Academy of Science, 3: 47–52*

• *Helm, Georg (1881),«Ueber die Vermittelung der Fernewirkungen durch den Aether», Annalen der Physik, 250 (9): 149–176,Bibcode:1881AnP...250..149H,It hurts:10.1002/andp.18812500912*

Isenkrahe, Caspar (1892),«Über die Rückführung der Schwere auf Absorption und die daraus abgeleiteten Gesetze», Abhandlungen zur Geschichte der Mathematik, 6, Leipzig, pp. 161–204

• Maxwell, James Clerk (1878), «*atom*», in: Baynes, *TS,Encyclopædia Britannica*, 3 9th ed. , New York: *Charles Scribner's Sons, pp. 36–49*

• *Maxwell, James Clerk(1878), «attraction», in: Baynes, TS,Encyclopædia Britannica, 3 9th ed. , New York: Charles Scribner's Sons, pp. 63–65*

• *Peck, JW (1903), «The Corpuscular Theories of Gravitation», Proceedings of the Royal Philosophical Society of Glasgow, 34: 17–44*

- *Poincaré, Henri (1914) [1908], «injury's theory», Science and Method, London, New York: Nelson & Sons, pp. 246–253*
- *Preston, Samuel Tolver (1895), «Comparative Review of some Dynamical Theories of Gravitation», Philosophical Magazine, 5th series, 39 (237): 145–159,It hurts:10.1080/14786449508620698*
- *Taylor, William Bower (1876), «Kinetic Theories of Gravitation», Smithsonian Report: 205–282*
- *Van Lunteren, F. (2002), «Nicolas Fatio de Duillier on the mechanical cause of Gravitation», in: Edwards, MR, Pushing Gravity: New Perspectives on Le Sage's Theory of Gravitation, Montreal: C. Roy Keys Inc. pp. 41–59*
- *Zehe, Horst (1980), «Die Gravitationstheorie des Nicolas Fatio de Duillier»,ISBN3-8067-0862-2, Hildesheim: Gerstenberg, Archive for History of Exact Sciences, 28(1): 1–23, Bibcode:1983AHES...28....1Z,It hurts:10.1007/BF00327787*
- *Zenneck, Jonathan(1903),«Gravitation»,ISBN978-3-663-15445-7, Encyklopädie der Mathematischen Wissenschaften mit Einschluss Ihrer Anwendungen, 5 (1): 25–67,It hurts:10.1007/978-3-663-16016-8_2 [dead link]*
- This page was last edited at 00:00 on July 28, 2021.
- This text is made available under the terms of the license Creative Commons Attribution-Share Alike 3.0 Unported (CC BY-SA 3.0); may be subject to additional conditions. For more details, see the conditions of use.[13]

The idea of ether dates back at least to antiquity, with Aristotle. For him, the ether would be a divine substance, which is not created, is indivisible and indestructible, in addition to being endowed with movement.

---

[13] https://en.wikipedia.org/wiki/Mechanical_explanations_of_gravitation

Descartes conceived of an ether made up of particles whose motion formed the different types of matter.

Newton came to believe in a static ether in which the density variation would be responsible for the gravitational effect. This allows us to imagine an ether in which the density would be greater in the farthest regions of the celestial bodies and smaller in the vicinity of these, so that the greater pressure in the regions of greater density would cause a force in the bodies, moving them towards each other. It is a fascinating explanation, like several other explanations based on some kind of ether presented in the past, despite the important criticisms they received, as shown above.

Huygens' luminiferous ether is intended to explain the propagation of light, conceived as a wave, as opposed to the Newtonian corpuscular conception of light.

Maxwell believed in an electromagnetic ether that propagates electricity, magnetism and light, conceived as an electromagnetic wave.

Currently, the theoretical Higgs field consists of an ether of Higgs particles, whose purpose is to explain how particles of matter acquire mass when moving through this ether.

The conception of an ether with static particles was apparently decisive in the conclusion reached by the prevailing scientific community in light of Michelson and Morley's results.

Some forms of ether idealized throughout history, even postulated today, do not involve the algorithmic principle, since the movement of its particles does not result in particles of matter, and some of these forms of ether are specifically designed to try to explain only gravitation, light or electromagnetism, but not the normal matter.

In summary, not every conception of ether aims to explain the constitution of matter and, thus, not every ether harbors the algorithmic principle, in the sense treated here.

The algorithmic principle of AMT

The basic difference of the algorithmic principle from the Algorithmic Matter Theory between other conceptions is its universal character, in two aspects. (1) It applies to the entire extent of the universe, supposedly infinite; (2) Applies to all size scales, assumed infinite.

I announce that according to this principle, all particles of matter and energy are formed by algorithms of movement of particles of an immediately smaller size scale. This is the fundamental principle, the basic postulate of the Algorithmic Matter Theory.

Particles of a dimensional scale (size scale) can be considered fundamental relative to that scale. In this way, the character of a fundamental particle is relative to a certain scale, since this fundamental particle is formed by the movement in a cyclic flow of particles of the immediately inferior dimensional scale, forming an infinite succession, both towards the inferior scales and superior scales in relation to the considered scale. This includes the human scale, both "up" and "down" in the hierarchy of scales.

This algorithmic principle breaks in a deeper way with the anthropocentrism that permeates the other physical models, which assume that the scale of the fundamental particles perceived by the human being is the lowest possible level, treating as absolute the fundamental character of the particles of matter and energy.

The algorithmic ether stratified on dimensional scales

Despite the apparently complicated name, it stems from the AMT algorithmic principle, which I will refer to only as the algorithmic principle, the existence of an algorithmic ether stratified in dimensional scales, which is nothing more than an ether that admits the vacuum formed in regions of the absence of fundamental particles of the considered scale. This algorithmic ether is stratified into dimensional scales, simply because in each of them the fundamental particles are formed by ether particles of the next smaller size scale. What is a particulate medium (ether) for one scale is a vacuum of fundamental particles of matter for the same scale – for example: empty space without particles like protons and electrons, though occupied by ether particles.

The shape of the particles

We have no information about the likely differences in the conditions of the cosmos at each dimensional scale, much less its evolution. We are still in its infancy to discover the evolution of cosmological conditions relevant to our scale, which ranges from subatomic particles to the entire observable universe, passing through large clusters of galaxies and what appears to us as colossal filaments of concentrated matter formed by galaxies. So, as for the other scales, we can only imagine the result of the evolution of the respective conditions in the multiplicity of forms of their particles and in the presence of possible dimensional subscales, in a certain cosmological time. As a speculative example, our entire known universe may be but a tiny aspect of a single particle on the cosmic scale immediately above.

We can only imagine the level of homogeneity and stability of particle shapes at scales different from ours. We can speculate that each dimensional scale can present a variation in ways of presenting matter and energy as great as the variation we find in our scale. Thus, planets, stars, galaxies, clusters, nebulae, molecules, atoms, electrons, protons, neutrons, quarks, photons, gluons, and every form of matter and energy that we distinguish on our scale may be nothing more than ether particles on the immediate scale above ours.

At certain scales, evolutionary conditions may not have allowed for as much variation, and the particles may resemble simple swirls of particles, unstructured in the form of exploding stars or rocky planets, unlike at the human scale. Likewise, galaxies on our scale may be sufficient to characterize small ether particles on the next scale up, which would also include large "clusters" and filaments of galaxy clusters. To corroborate this possibility, I remember that on some day in the long years that I have consumed to write and rewrite this book, I photographed a very interesting example of the formation of filaments. If we add the proper amount of instant coffee to milk heated to a suitable temperature and stir until a visually homogeneous liquid is obtained, after some time the mixture has rested, and beautiful whitish filaments are produced. Has anyone noticed this?

Motion Algorithms

In the current state of knowledge, a candidate for a simpler algorithm, forming the most elementary particles of each dimensional scale, is the cyclonic, vorticiform algorithm, very common in nature, at least in the human dimensional scale, as in the movement that originates planetary systems, in the accretion disks of black holes, in the configuration of spiral galaxies and the absorption of one star by another, in binary systems. Let's take it as a starting point for "algorithmic reasoning".

If we could generalize about the fundamental, elemental nature of all physical phenomena, then we would say that the fundamental unit of the physical world is algorithms of motion. But what can we understand as motion algorithms? Motion algorithms are a simplified way of referring to *natural* patterns of processing the movement of particles forming sustainable flow cellular structures – under certain conditions – which form particles of larger scales, in a fluid cosmic medium stratified in successive dimensional scales and from which all known matter and energy depend on their constitution and displacement in space and time. This view defines the known world as a *universe of fluid particles* in all its dimensional scales.

The term algorithm is used to convey the idea of a natural pattern of processing the translational movement of particles of a given dimensional scale, probably rotary translation movement in the simplest cases, determining a flow forming the elementary particles of the immediately superior dimensional scale. It is important to bear in mind that each kind of motion algorithm may one day be described with the aid of computational and mathematical tools that take into account characteristics that may be associated with the interaction between particles of a given dimensional scale, which we can call horizontal interaction. If we consider the comets, the Sun, the Earth, and the Moon as particles, we will call their gravitational interaction the horizontal interaction, because they belong to the same size scale. The term horizontal here does not refer to a line or a special plane of interaction, but to the interaction between particles of the same dimensional scale, in this case, the human scale, which, as we have already discussed, ranges from subatomic particles to clusters or superclusters (galaxy clusters). Eventually, the effects that particles of other scales may cause in the algorithms of the considered scale should also be taken into account, effects which we can call vertical interaction.

General Implications of AMT

The Algorithmic Theory is still in its descriptive phase in the verbal, explanatory way, not dealing with any quantitative prediction yet. We are still trying to work our minds to reach a consistent description of the nature of physical phenomena, before trying to use any mathematical, quantitative instrument. This has happened many times in the history of humanity and physics. We can cite as great examples Aristarchus' explanation for eclipses, despite his eventual quantitative calculations, the explanation of Copernicus for the movement of other planets relative to Earth in the Solar System and the explanation of Einstein for the Brownian movement of particles of dust in a liquid. The mathematical formalization of every theory, as a condition of its acceptance, is the method that has predominated in theoretical physics for many decades – and that seems to be intrinsically inscribed in the current concept of physics as a science. I am not afraid to believe, however, that like other crucial moments in the evolution of science in general, a new explanatory model must once again precede a mathematical formalism compatible with it. On a few historical occasions, a new speculative model has been as necessary as it is today to inspire and induce the creation of a new path to scientific production.

The previous observation was made to make it clear that, according to AMT, there are natural algorithms for processing particle motion that are independent of their association with any geometric or statistical model. In other words, these algorithms (movement patterns) are present in the intricacies of matter and the cosmos without their nature being confused with any mathematical construction discovered or invented by human beings to describe their behavior. Nevertheless, the concept of the algorithm discussed here involves the notion of flow patterns, with different architectures and varying degrees of stability, determined by the evolutionary conditions of the cosmic environment that sustains matter and energy. This relative regularity, and especially the conditions that provide extraordinary stability for the most common particles on a given dimensional scale, as in the case of electrons and photons on our scale, imply the possibility that the algorithms involved will be described mathematically, although this does not seem a trivial task for someone who does not deal with the mathematical model building as I do.

This mathematical independence is especially opposed to the idea that the geometry of General Relativity, although extraordinarily useful in the macroscopic description of gravitational phenomena, could account for the nature of space and time. Likewise, such independence opposes the idea that quantum statistics, equally useful in particle physics, is capable of accounting for the nature of matter particles. We thus oppose the idea spread by Einstein that the nature of space – associated with time – is geometric, and the idea spread by Quantum Mechanics that the nature of elementary particles of matter is statistical and that the specific behavior of an elementary particle, such as trajectory and linear momentum, depends on the particle being observed by humans through detectors designed and built by us. AMT definitely does not sit well with the Copenhagen interpretation of quantum mechanics.

It seems appropriate to assume that we share Einstein's opposition to the most radical explanatory versions of the uncertainty principle of quantum mechanics, either in the notion that a particle of matter or energy assumes no specific place in a space-time binomial except at the moment it is detected, or in the idea that a particle could be located at the same time in infinite space coordinates, although with varied degree of probability, besides being able to travel simultaneously through infinite alternative trajectories between two points in space, as long as it is not subject to any type of observation, as understood by Feynman.[14]

We believe that the technology applied in the observation of the movement of an atomic or subatomic particle, that is, for the determination of its location, velocity or spin, alters some of these defining variables of the state of the particle, which leads us to admit the uncertainty principle, in a less drastic version, reduced to the limitations of our technological capacity to observe the state of the so-called elementary particles without interfering in the result of the observation. AMT, on the other hand, does not seem to be incompatible with the idea of quantum foam, in the sense that, under certain conditions, particles are created and annihilated (or not), provided that this arises from algorithmic interactions between particles of inferior scales and not from the oscillation of the geometric space-time fabric supposed by General Relativity and incorporated by Quantum Mechanics and String Theory.

Energy quantization

---

[14] GREENE, 2001, pp. 130-131

The AMT does not object to the idea of quantization of energy, which seems to us to be sufficiently demonstrated by current physics. If what I am going to mention goes unnoticed by the reader, quantization is easily explained and even emerges from AMT, when we realize that a stable pattern of the algorithmic flow defining a particle can perfectly be seen as a quantum of energy. This leads us to opt for the nature of light as a set of non-corpuscular particles, but algorithmic, in the sense we are proposing. Eventually, groups of particles that form light can move in space maintaining their relative positions in order to form packages with a determined number of particles arranged in space in the form of an Indian line, in a straight line, or forming transverse waves of particles. Alberto Mesquita Filho speculated about the formation of a straight Indian line[15], but I add the possibility of a wave-shaped queue, sinusoidal, more for aesthetic reasons, and as a slightly more attractive hypothesis in the discussion of the hypothetical duality of the (undulatory and corpuscular) nature of light, currently accepted. But the Indian queue proposal is also very interesting. We accept Alberto Mesquita Filho's argument that the normally observed unitary quantum effect does not necessarily depend on the presence of indivisible corpuscles, but eventually on a set of particles forming a single package. For the AMT, such particles can maintain their relative spatial arrangement by inertia or by some active algorithmic mechanism between them, in the formation of the quantum packet, although we should not discard a quantum as a cellular algorithm formed by a solitary nucleus of flux. In this way, a quantum of light can be a simple elementary algorithm or consist of a more complex algorithmic one, formed by the combination of smaller flux cores, algorithmically interconnected in their displacement or just inertially articulated.

Still on the notion of motion algorithms

To try to clarify the notion of motion algorithms, in the context of AMT, we can resort to the following analogy. We have already seen that the term algorithm refers to a certain pattern of cellular particle flow, sustained by the conditions of the universe present at a given moment in a region of space, especially the natural conditions of the cosmic particle

---

[15] https://www.ecientificocultural.com/ECC3/polar00.htm

medium proposed by the theory. An example of a very stable motion algorithm by human standards of observation, which forms what we call a supercell, is the great red spot on Jupiter, which we know to be formed by an immense hurricane on the gaseous surface of the planet. That vortex lasts for centuries, who knows for millennia, a fact that attests to its considerable stability. As long as the conditions for its support prevail, the spot will continue to fascinate both amateur and professional astronomers. The great spot of Jupiter is a typical example of a movement algorithm, in the concept adopted by the AMT. Analogously, we propose that the algorithmic nature of the large spot may be similar to the nature of an electron, for example. While the big spot consists of the sustainable rotational movement of gas molecules, the electron may be the result of the rotational movement of particles of a sustaining matter, nothing preventing the electron algorithm from being a much more complex flow structure.

To close this section on the concept of the algorithm of motion we can summarize it by saying that it is just a natural pattern of particle movement, that is, a flow of particles that eventually form a supercell or any other flow structure, capable of assuming the behavior of a particle in a higher dimensional scale, adding that, in the model proposed by TMA, different algorithms for different elementary particles can interact and form combined and more complex algorithms, as in the case of protons and neutrons, which participate in stable interactions in which particles of theoretical strength,, with existence corroborated by images of collisions in particle accelerators, such as gluons, act in the aggregation of groups of quarks. In the structure of atoms, protons, neutrons and electrons coalesce together with some particles of force. Each algorithm (particle) has the natural ability to absorb or release energy in the interaction with other algorithms, elementary or not, and even dissipate them (destroy them, absorbing them) or release them (constitute them), as in the case of the interaction between a photon and an electron. Ultimately, if we consider the countless possibilities of interaction between the algorithms that form different kinds of matter and energy particles, including the mutual annihilation between matter and antimatter, we can conceive of the entire universe as a single, complex and constantly evolving algorithm, responsible for processing the movement of particles of different dimensional scales, bearer of several facets, that particularly considered reveal the different particles that inhabit the various levels of cosmic existence.

Dark matter

AMT proposes that the algorithmic ether, which we can also call sustaining matter, is nothing less than the medium that allows the formation of what physics today refers to as dark matter, responsible for the additional gravitational effect believed to be present in the cohesion of galaxy objects. Likewise, this same sustaining matter, dark matter, would be responsible for the formation and propagation of photons, (by means of algorithms that differ from the electronic algorithm) and other fundamental particles of our human scale.

Algorithmic matter

I considered it important to incorporate the term algorithmic matter into the name adopted for the theory to symbolize the premise that matter is not a physical reality independent of a cosmic medium. In this model, the matter would not exist in an absolute vacuum, devoid of sustaining matter. On the contrary, its existence is intrinsically linked to the movement of particles present in a particulate medium. What physics today considers to be elementary particles of matter, such as electrons and quarks, are, for AMT, just the result of the continuous movement of much smaller particles. In this new microscopic context, space-time geometry is replaced by kinetic interactions, observing that elementary particles are neither continuous nor indestructible corpuscles. The particles considered to be elementary on a given scale are formed by the motion of perhaps many billions or trillions (these numerical expressions were chosen at random) of particles of a smaller scale. This is how energy condenses into matter, according to Algorithmic Theory. The kinetic energy of potentially billions or trillions of dark matter particles accounts for constituting and sustaining the ordinary matter particles and their respective characteristic movements, including rotation (if any) and translation. The trajectory, the moment, and other variables associated with matter particles no longer depend on observation. Some characteristics of particles of a given species may depend on a possible rotating component of the respective algorithm. The translation motion of these particles along larger structures such as atoms, on the other hand, depends on the interaction between the individual algorithms of the structure under consideration and between these and the particles in the sustaining medium.

Cosmic choreography

String Theory proposes that the universe works like a symphony, through the musical vibration of the strings that constitute the elementary units of matter. It also refers to a supposed dance of these strings. The Algorithmic Theory, in turn, radicalizes this last image, attributing maximum importance to the choreography between particles of different scalar levels. All physical phenomena, for AMT, derive directly from the choreography of known matter particles (normal matter) and sustaining matter particles (dark matter or algorithmic ether). We are categorically stating that, according to the scheme of the Algorithmic Theory, matter, energy, gravity, inertia, electromagnetism and the phenomena derived from them, including time, all result from the choreography of particles in a sustaining medium stratified in successive and infinite dimensional scales.

Properties of the algorithmic ether not discussed by Einstein

When Einstein managed to bury, almost completely, the idea of an ether of any nature, one of his arguments was that a cosmic medium formed by particles responsible for the undulation of light would offer resistance to the movement of the stars, slowing it down, which would be incompatible with Galileo's law of inertia. Einstein refers to several models of ether, all of which failed, according to him. Perhaps we do not know all the conceptions of ether proposed until the innovative Einsteinian prediction that electromagnetic waves would simply dispense with a medium for their propagation, but apparently, the versions in vogue at the end of the 19th century departed from the premise that the particles of the luminiferous ether of Huygens and Maxwell's electromagnetic ether would have little or nothing to do with the constitution of matter particles.

Cartesian ether and AMT ether

The premise of AMT resumes the Cartesian principle that the ether, more than interacting with matter, constitutes its raw material. In this way, the ether postulated by AMT cannot offer resistance to the movement of the matter of celestial bodies, since the existence of this matter and its displacement in space fundamentally depend on the existence and movement of the particles of this ether, which makes the aforementioned Einstein's argument invalid against the existence of the algorithmic ethereal medium conceived within AMT. On the other hand, despite being compatible with some important aspects of Cartesian physics, the ether we postulate presents fundamental distinctive features concerning that imagined by Descartes. The Cartesian ether is formed by corporeal primordial particles, rigid, capable of being broken, endlessly worn and shaped by each other by direct contact in their eternal movement, without the possibility of empty spaces between them. The algorithmic principle that all physical phenomena result from the incessant movement of particles of a substrate that permeates the entire universe, was already present in Cartesian physics, but the Algorithmic Theory that we are proposing presents a different algorithmic principle, based on a substrate with essentially distinct structure and mechanics, despite some similar notions, such as the probable importance of vorticiform algorithms. We understand that a special criticism should be made of Descartes' particulate ether. For him, the particles of ether were created without empty space between them and set in motion by God. After the initial impulse, the particles were breaking and wearing out, remaining without interstices. We claim that this is completely impossible since for them to move relative to each other, and even for them to crack and break apart, a minimum of empty space between the particles would necessarily be formed or required.

The gravitational phenomenon

This must be the high point of any theory that intends to unify or replace paradigms present in General Relativity and Quantum Mechanics: to explain what gravity is, consistently from the macro and microscopic points of view. As already mentioned, quantum field theory (quantum gravity) has been striving in this direction, through a model completely incompatible with ours. In the case of AMT, this is not exactly a unification between the two great areas of physics, because by denying the physical existence of the space-time fabric, we are denying an essential aspect of both currently dominant theories. The conceptual aspect that we are opposing is an important part of the obstacles that irremediably prevent the unification of both non-algorithmic theories. If we removed the concept of the space-time continuum and its geometric character from the Einsteinian scheme, the conceptual problem between the two theories would diminish. The same would happen if we removed the uncertainty principle and the statistical character of the fundamental particles from Quantum Mechanics. In both cases, we would amputate fundamental concepts from those theories without which they would not survive, despite the beauty and successful predictions of their mathematical formulations. We reiterate that both theories lack a physical substrate, the "modus operandi" of nature. Mathematical representations of the world that represent themselves – since the world is confused with the representation – are not satisfactory representations of the world.

As is known, in the scheme of General Relativity, gravity is incorporated into the fabric of space-time. It ceases to be a force of attraction at a distance caused by bodies of matter, in the conception often attributed to Newton (perhaps wrongly), and becomes a result of the curvature that bodies endowed with a mass imprint on the space-time continuum. If before, by interpretations attributed to Newton's scheme, gravity was a force of direct attraction, at a distance, between bodies with mass – although it is said that for Newton himself the existence of an eventually immaterial agent that transmitted this force was indispensable –, since Einstein, gravity has become the effect, on material bodies, of the distortion caused by these same bodies in the fabric of space-time. In General Relativity, space-time takes on the role of the so-called intermediate entity, the messenger of gravity, apparently predicted by Newton.

On the other hand, the current physics of known minuscule elementary particles of matter and energy offers a model according to which, due to the uncertainty principle, on an ultramicroscopic scale everything seems to oscillate frantically, including the gravitational field, an attribute of the space-time fabric, independently of the previous existence of matter in the vicinity of the cosmic region considered. For Quantum Mechanics, even without the prior presence of atomic matter in a certain portion of space, or in its relatively close surroundings, space itself oscillates and produces dramatic gravitational variations, on a very small scale, constantly creating its gravitational field and its opposing particles of matter or force, which mutually annihilate each other. This intense microscopic oscillation of the gravitational field is not in line with the soft curves of Riemannian geometry that describe gravitational fields in General Relativity.[16] For this reason, among others, basically due to the opposition between the smoothness of General Relativity and the strong quantum oscillation, concerning gravity, the two theories clash and require unification, reformulation, or replacement by a more comprehensive model. But, we repeat, this only occurs because both theories adopt as a premise a space-time continuum, to which gravity is incorporated as a result of its smooth macroscopic deformation or its frantic microscopic oscillation.

In the AMT scheme, all this goes away. Gravity is not the result of the curvature of a continuous cosmic fabric. This continuous fabric does not exist! Gravity becomes the result of the movement of particles in the dark matter ether or algorithmic ether. The starting point for understanding gravity in our scheme is the algorithmic particles of matter. Each particle is the result of the natural mobilization of the algorithmic ether. These elementary particles interact with each other and associate to form material bodies. The bodies are discontinuous, present interstices and the ether is mobilized in these interstices, either to form matter particles or a secondary mobilization in interstices, as a byproduct of material algorithms. We need to imagine how this secondary mobilization can be, in the interstices of the matter of a material body.

In the first hypothesis, in this scheme, a macro flux of algorithmic ether particles is formed by the set of microfluxes of a material body and converges to the center of this body, with increasing intensity towards the center. In this hypothesis this macro flux is predominantly ordered, such as a smooth flow of water toward a sink, that is, the ether particles are pulled by a large material body, from a certain distance (just as the water molecules are brought gently by the sink in cases where they do not form whirlpools), taking with it other smaller material bodies (similar to what happens with a paper boat on the surface of the water that is being drawn

---

[16] GREENE, 2001, pp. 138-142

in, or a chlorine ion below the surface, or even a fish swimming in a tank unaware that it is being swallowed with the water). This first model holds some resemblance to Bernhard Riemann's model, where bodies with mass are sinks of an incompressible ether. In our first hypothesis, the condition of incompressibility is not necessary.

In a second hypothesis, in the surroundings and the meanders of a massive body, the secondary mobilization of the ether, provoked by the algorithms of matter, occurs in the form of a predominantly chaotic movement, that is, the particles of the ether do not move in an orderly way, they do not follow a preferential line towards the center of mass of the body. In this second scenario, the ether particles follow chaotic trajectories, interacting with each other, probably with interactions between scales (which we call vertical) influencing their movement, in which case the degree of freedom of movement of the ether particles is very large, compared to the first hypothesis. While in the first hypothesis, the particles of the ether are subject to a strong limitation in their relative motion, like the molecules of water, in the second scenario the secondary motion of ether particles more closely resembles the motion of house dust suspended in the air, shimmering in a beam of sunlight. Although chaotic, the second scenario assumes a gradient in the chaotic agitation of the ether particles, which increases towards the center of mass of the material body, both inside and outside the boundaries of this body.

A third scenario combines the two previous types of secondary ether mobilization, smooth and chaotic. The predominance of one or the other may depend on some variables such as the distance to the massive body, the fact that the region considered is located inside or outside the limits of the body, the distance of the ether particles to the algorithmic nuclei of the material particles within the limits of the massive body, among others. As an image, I remember that once, in childhood, I dragged a slipper in the water at the beach, in a vertical position, and I found it interesting how the slipper attracted a small red plastic ball, which floated on the surface, despite the great turbulence of water that formed behind the slipper.

In any of the three scenarios, the basic idea is the same: the set of structured microflows that form the massive body causes some kind of secondary mobilization that presents an intensity gradient towards the center of the body, and it is this secondary mobilization that causes the effect gravitational.

The smooth-flow model has the disadvantage of the difficulty of explaining where the macro flux particles go after going through the algorithms of the material particles of the body and reaching the center of the massive object. This explanation is more intuitive in the chaotic secondary motion model, with an increasing agitation gradient towards the center of the body. The particles are simply not consumed, remaining in agitation in the meanders of the massive body and on its periphery. Another disadvantage of the smooth flow model is the difficulty of offering an acceptable visualization of the macro flow, from the most distant regions, even without the interference of other equally massive bodies nearby, when we think of the inertial translational movement of the body. It would be like trying to imagine the flow lines in a huge water reservoir, whose open drain is a constant displaced. Nevertheless, logic seems to indicate that the secondary movement presupposes an algorithmic ether with a degree of freedom that can vary according to the local flow conditions, allowing order and chaos to be combined in the exact measure determined by these conditions.

The Algorithmic Theory emerged from the attempt to explain gravity. So perhaps the best way to explain the algorithmic gravitational scheme is to tell the history of the theory. Not the history of previous ideas produced by humanity compatible with it, but the history of my personal reflections that motivated its formulation.

CHAPTER III
THE BEGINNINGS OF THEORY

The beginnings of theory

The present theoretical proposal began to take shape from an unavoidable reluctance that I always had in accepting that something like a vacuum could bend. The idea that something can bend empty space has always struck me as intuitively illogical. Even more so if this curvature could alter the motion of planets and stars, or objects like a simple apple thrown in any direction above the Earth's surface. Would curvatures of empty space be able to accelerate an asteroid, trap air and water to Earth and deflect light strongly enough to trap it in a black hole or simply alter its trajectory? How could the vacuum, which in essence should designate nothingness, contort itself so intensely, dragging everything around it, the stars, the planets, the gases, even the light? These questions full of incredulity never left me since I became aware of the concept of curvature of space, without even knowing that the postulated curvature would have an essentially geometric nature until the day I came into contact with the possible existence of dark matter and its probable gravitational properties.

Attractive oscillation

Upon learning about dark matter, the thought that almost automatically came over me was that somehow dark matter would be attracted to normal matter. A new question then arose. What could dark matter "see" or "feel" in the normal matter that would attract it? In other words, what is there in normal matter, at its core, that is capable of attracting dark matter? When thinking of the entrails of ordinary matter, I thought of its inner movement. The oscillatory movement of its molecules caused by its latent heat. It was the beginning of an outline. The agitation movement of normal matter molecules seemed to me, at that moment, the only thing capable of attracting dark matter. I imagined that this could be enough to generate a flow of dark matter toward a planet, for example. This dark matter would be formed, in this first draft, by particles different from normal matter particles. Once the flow is formed, this could be the element responsible for a planet's gravity. The greater the mass of the planet, the more vibratory molecular motion it would have and, thus, the greater its ability to accelerate dark matter against itself. This would explain the relationship between mass and the force of gravity. I began to get excited about the possibility of neglecting the gravitational curvature of empty space.

Flow bombing

For this, dark matter should have some special properties, such as the power to influence, through an intense bombardment of its particle flow, the movement of another body, say, an asteroid, the trajectory of which crossed the attractive flow caused by the planet. This other body, in turn, due to its internal molecular movement, would also cause a flow of attraction of dark matter, with an intensity proportional to its mass. This second flux, caused by the asteroid, would also influence the movement of the planet that attracted it, as it would be a property of dark matter, in flux, to modify the movement of matter bodies. Thus, each material body would cause its own dark matter attraction flow, and each flow would be able to attract the movement of other bodies, with their respective flows, in proportion to its mass. Obviously, this flow would lose intensity in the regions farthest from the material body, consequently decreasing the power of attraction between two bodies with the increase in their relative distance. A surprisingly simple explanation for macroscopic gravity. But would this also apply to microscopic gravitation? Keep in mind that these were just initial speculations of an idea that would evolve a lot. Incidentally, the idea would be replaced.

Annihilation of dark matter

Molecular agitation, in this initial model, would attract dark matter particles. But what would happen microscopically when a dark matter particle approached a stirring molecule? I was forced to imagine that somehow this churning would consume the dark matter particle. Successively the dark matter particles would be consumed, and annihilated, one after another, which would decrease the density of the medium formed by them in the vicinity of the normal matter particles and would force a displacement of the neighboring dark matter particles towards the atomic matter molecules, thus sustaining the gravitational flux of dark matter. At this stage, Algorithmic Theory still did not have this name and I had not even realized that I had just imagined my first motion algorithm model, extremely rudimentary, but which to me sounded like the carrier of the fundamental principle of a new physical and philosophical conception of the universe. In fact, this idea was not new, if we consider the efforts of Descartes and other defenders, some of them current, of the idea that the gravitational effect is caused by the flow of particles of some kind of ether.

Dark matter and the fate of the universe

These reflections occurred to me, if my memory serves me correctly, during the year 2010. I had these ideas in mind for a few months, and a certain concern about the destiny of the universe arose. If normal matter consumed dark matter and if this consumption were responsible for generating the flow of gravitational attraction, this process would consume dark matter at very high rates and it would not be by chance that the most distant galaxies would be distancing themselves from each other in such a way accelerated form, surpassing, in its removal, the speed of light. Assuming that dark matter could be finite in our universe, and that, in the process of cosmic expansion, galaxies, and their clusters retain and consume dark matter, attracting it in flux, as they accelerate away from each other, it would be natural to conclude that the density of dark matter would progressively decrease in the regions of the universe that are emptier of material bodies. This could form dark matter bubbles around galaxies with increasing dark matter densities towards the center of each bubble, that is, towards the center of each galaxy, tending to form increasingly empty regions, with lower dark matter densities in the space between the galaxies. What would happen when dark matter runs out after billions and billions of years of consumption? Simply normal matter in galaxies could disintegrate, without the converging pressure of dark matter, and that would be the end of the universe as we know it. And what would dark energy be, currently considered responsible for the accelerated separation of galaxies? At that time, I speculated that it could just be the absence of dark matter, or even its rarefaction, an idea I still haven't completely abandoned.

What are we made of?

This model, still incipient, remained dormant for more or less a year until at the end of June 2011 the theory would take a quality leap. It was on the day that I commented to my work colleagues, in a bureaucratic public office, that I found it surprising that most people reach our age, in their fifties, without knowing that, according to General Relativity, in the vicinity of a planet like Earth, because of its great mass, time runs more slowly than in regions of space a little further away, such as on the surface of the Moon, which has a smaller mass – this without considering the difference in speeds on the surfaces of the planets. For almost all the colleagues present in the room, this was a revelation. They did not know about this phenomenon, taken as a fact by science. As the conversation goes back and forth, I ended up commenting that I was trying to develop a theory, which, at the time, it was already kind of forgotten, about the relationship between gravity and dark matter. Incidentally, dark matter almost nobody in the group had heard of. I took the opportunity to advertise that, in the evening of that same day, an episode of the series "Through the Wormhole", with Morgan Freeman, would be shown, produced by Discovery Channel. It would be my first time watching the series. The episode titled "What Are We Really Made Of?" addressed the scientific enigma of atomic cohesion and, in the end, discussed the nature of elementary particles of matter such as electrons and quarks, defining them as portions of condensed energy. During the program, I tested for the first time, in the light of a scientific exposition of high didactic level, the theory of the flow of dark matter, without finding anything that could refute it, which got me really excited. On the contrary, contact with models such as the Higgs field made the gravitational flow of dark matter appear more and more true. But so many revelations about the nature of matter and about the gaps that still exist in this area of physics soon gave me the feeling that something very elementary still needed to be conceived and understood.

Condensed energy

Freeman's way of referring to the electron, as energy condensed into matter, made me defiantly uncomfortable. I wondered what that could mean. The narrative highlighted that atoms are no longer seen as those solid and continuous balls of matter that were imagined in the past. They started to be made up of smaller particles, and Freeman was quite emphatic when he said that quarks and electrons are not like continuous, solid balls either, being more like portions of condensed energy – and not so condensed. He made it clear that if we could look at the fundamental particles that make up the things we consider solid, we might only see tiny clouds of energy. I don't remember if these were his exact words, but that's how I registered the idea. The depth of this idea struck me as overwhelming. What could the notion of condensed energy in matter mean in terms of physical reality? As it turns out, Einstein quantitatively demonstrated the mutual convertibility between matter and energy, but, in the qualitative, descriptive aspect, how could we imagine the mechanism responsible for this mutual transformation?

Crossing the border

So, I tried to identify what could be fundamental in matter and energy that would allow them to cross the border between one thing and another without giving up their essential and common nature. The answer I found was fundamentally identical to the one I had found for the question I had asked a few months earlier, about what could be in the normal matter that attracted dark matter: the movement!

The movement

We all have the notion that everything is in motion in the universe. The stars, the oceans, the air, and even an iron bar resting on a table have an enormous amount of movement in its molecules, atoms, and electrons. Nothing is stopped. Even if two objects maintain the same relative positions to each other, they will not be stationary concerning other objects in the cosmos. And even these two objects will not be able to maintain their relative positions forever, as everything in the world changes. Furthermore, if they are two real-world objects, they are likely to be made up of molecules and atoms that vibrate, not just internally, but also relative to the molecules and atoms of the other object that forms the pair.

Matter and energy

Returning to the question of matter and energy, if we admit that *movement* is what both have as essential and common, we need to reflect on how the fundamental particles of matter and energy can *always* be in motion. How could energy in motion condense into matter? Could this condensation mean that in matter the movement is more concentrated, while in pure energy it is more dispersed? Could matter be energy in a more concentrated form of movement? Movement of *what*?

Amid these questions, I remembered the medium formed by dark matter particles that I had imagined earlier. And then came the most fundamental idea of what would become the Algorithmic Matter Theory. The hypothesis occurred to me that the natural energy contained in the dark matter medium could manifest itself in the form of a kind of whirlpool of those unknown particles. Each whirlpool could form a cell of concentration of energy in rotation, units that could constitute the particles known as the most elementary of common matter: electrons and quarks.

The picture was almost complete. Energy would be transmitted through space by the movement of dark matter particles, and the atomic matter would also be the result of a specific form of movement of the same particles of unknown matter. There would be a basic reality common to matter and energy, intrinsic to both, which would be the dark matter particle in constant movement. The difference between ordinary matter and energy would be dictated only by the specific way dark matter particles move. By concentrating motion, we would condense energy into particles of matter; by dispersing it, we would convert matter into energy. The fundamental unit of matter would no longer be an independent particle, but a parade of much smaller particles arriving, participating in a choreographed movement, bowing to the judges of the artistic presentation and giving way to those who have not yet arrived for the show.

Movement of what?

The next question would be: if the particles of normal matter are the result of the movement of smaller particles, in a sustainable flow unit, with its architecture, what would the sustaining particles of this flow be made of?

The basic and intuitive way in which one particle transmits motion to another is the direct collision between them. Could there be another way? Who knows the attraction at a distance, imagined by Newton's followers to explain gravity? In this case, what would be the transmission agent? Could it be the curved deformation that a particle causes in space, causing another particle to slide down a kind of cosmic toboggan, like what Einstein imagined for massive bodies? Discarding such possibilities, I assumed that nature could reserve only one basic way to perpetuate movement and that this was through direct contact between particles. Even if the particles were not solid and could take the form of any kind of cloud, energy field, or vortex, I imagined that only the approximation of a particle until reaching the border of another, as if it wanted to cross it, could significantly modify the state of motion of both. Would this idea be correct? Are there other ways for one particle to influence the motion of another?

When thinking about the condensation of energy in matter, I decided to explore the hypothesis of the transmission of movement exclusively by direct contact between particles. I would later learn that René Descartes had proposed this philosophical principle in the first half of the 17th century. Even not knowing that it was a Cartesian thought, this idea had nothing original, from my point of view, since I had just watched a television program that reported the certainty, spread by Quantum Mechanics, that the fundamental forces of nature are transmitted through particles, such as the photon (responsible for the transmission of electromagnetism), the gluon (which is responsible for transmitting the strong force), the weak force bosons and the gravitons. These are the so-called transmitter particles of the four fundamental forces of nature (among such particles, gravitons, transmitters of the force of gravity, are considered the only messenger particles of force that have not yet been experimentally confirmed, even if indirectly). Thus, the hypothesis of motion transmission exclusively through direct contact between particles seemed credible and natural to me.

Combining this thought with the idea of the flow of dark matter, now also forming microscopic vortices of matter, the chain of reasoning led me to the assumption that, in a given flow, whether straight or rotating, perhaps not all particles of the same size scale need to collide with each other all the time to maintain flow. For each scale of size considered, there could always be a medium, formed by smaller particles, which would be responsible for transmitting the movement of the flow to the larger particles, like a stream of water carrying debris. The flow could be maintained by direct contact between the particles of the medium – very tiny – and the particles of a larger scale, dragged by this medium. These smaller particles, in turn, maybe don't need to be in constant direct contact with each other to keep the flow going. It would be possible that their movement was transmitted by the collision, with them, of even smaller particles, often causing, if we could see these particles, the false impression that the movement of a particle of a certain size scale can directly influence and at a distance the movement of another particle of the same scale, without a direct collision between the two particles. Thus, a model began to be created in which the movement at a distance transmitted between similar particles would, in fact, be caused by the contact of much smaller particles, messengers of this movement, and so on, since smaller particles would need, as a rule, of even smaller messengers to communicate movement between them. The notion of a dimensionally stratified medium of particles was created, in which each elementary particle would be constituted by the rotating, cyclical movement of particles of a smaller dimensional scale, and so on. This basic model can support a series of variations, in the sense that some elementary particles, not necessarily on all scales, can be constituted by complex structures formed by simpler flow units, subject to a fluid coupling.

Whether this stratification is infinite or finds a universal limit, it would still have to be responded. But this issue would not prevent stratification from existing for countless dimensional levels.

Continuing with reasoning, if the most elementary particles of common matter can have a vorticiform architecture, as can be the case of electrons or photons, what about not-so-elementary particles such as protons, neutrons and atoms? This chain of thoughts occurred very quickly, from the moment the closing credits of the program presented by Morgan Freeman began to play on the screen.

On that last Friday of June 2011, the theory of the gravitational flow of dark matter attracted by the movement of molecular agitation underwent a revolution and ceased to be just a gravitational theory to also become a theory about the nature of matter and how energy and matter transform into each other. In the case of matter-to-energy transformation, if we could destroy the pattern of motion of a rotating flow cell that eventually corresponded to an electron, there would likely remain a wave or other form of kinetic energy dispersion through the dark matter ether. The initial idea that molecular agitation would be the factor responsible for the attraction of dark matter gave way to a continuous rotating flow of particles in the cosmic medium, at the same time forming the simplest particles of atomic (and subatomic) matter and causing gravitational flux. The gravity of celestial bodies and the gravity of microscopic particles were unified, supported by the concept of dimensionally stratified sustaining ether. From then on the model became richer and more complete and all I had to do would be to test it and improve it conceptually in the light of what I could find in more detail concerning the postulates about the nature of the universe according to the current theoretical models. The idea of motion algorithms (by that name) would come naturally in the days that followed, perhaps from reflections on how electrons and quarks might interact to form atoms. The result came in the form of natural patterns of interaction between the movements of those elementary particles. The elementary particles themselves would be natural motion algorithms (patterns) of the supporting particles, which individually would also consist of motion algorithms. The algorithms of quarks, when interacting with each other, would form protons, and neutrons. These, when interacting with the algorithms of the electrons, would be able to keep them attached to the dynamic architecture of the atom. Much later, already in mid 2012, with the continuation of reading about quantum theories referring to the cohesion of matter, especially regarding the role initially attributed to the Pi meson in the cohesion of the atomic nucleus, I began to realize that the force transmission particles are considered as transient (without existence independent of the atoms) and that this is also compatible with the algorithmic model.

But what happened to the idea of a continual flow of dark matter towards material bodies? And what happened to the idea of particle bombardment as causing the gravitational effect? We will see later that these ideas were replaced. Later, it will become clear that the continuous flow has been replaced by a gradient of chaotic motion in the interior and around bodies with mass.

The very idea that a particle remains in flow because of direct shock with particles of the same scale or lower scale has also been abandoned due to the fluid nature of the particles. But the idea of the necessary shock eventually led to the conception that each particle is formed by the cyclic movement of particles of the immediately lower scale, and so on, in infinite succession.

CHAPTER IV
ELEMENTS THAT EMERGE FROM THE THEORY

Algorithmic electron

We postulate that elementary particles of ordinary matter have the nature of a constant cyclic flow of particles of sustaining matter (dark matter), similar to some kind of vortex or another more complex cyclic algorithm. Quantum Mechanics postulates that electrons are endowed with a quantum state called spin, which may correspond, for us, to the revolving movement of these particles around a proper axis. There is also speculation among scientists that each electron may behave like a rotating magnetic dipole. Be that as it may, we propose, in the context of AMT, that the electron is constituted by a continuous cycle flow of particles of sustaining matter. The exact algorithm of this flow is yet to be modeled, but we can imagine that this algorithm involves, in addition to the rotating flow, in its equatorial plane, also a buoyancy flow. In this case, each electron could be compared to the flow of a small turbine. This type of combined spin and thrust flow is found in our atmosphere in the formation of hurricanes. I have not studied the structure of hurricanes, but I imagine, quite simplistically, that in addition to the flow lines representing the rotation of the cell, there are also the lines that suck in the air molecules underneath the rotating flow, sucking them from the bottom, then throwing them upwards, forming an upward flow. In my imagination, the rising air molecules, reaching higher atmospheric layers, spread out above the rotating horizontal plane flow. At the same time, a downward airflow forms on the opposite side. The upward flow then bends forming flux lines similar in design to Earth's magnetic field lines. The upward rotating flow around the eye of the hurricane bends radially over the top of the cell, eventually turning downwards at the periphery of the horizontal rotating flow until it again passes underneath the cell and returns to the upward phase in the central region of the vortex. If a hurricane works more or less like this, it turns out that the total flow of the phenomenon is formed not only by rotating components in the horizontal plane, but also by rotating components in the vertical planes radially arranged from the vertical axis of the center of the hurricane's eye, as with the magnetic field lines induced by an electric current in a circular wire, regardless of whether the representation of the magnetic flux direction eventually has the north pole as the input and the South Pole as the output of the flow, contrary to our hypothetical representation of the flow direction of a hurricane. This careless comparison is only intended to demonstrate that the algorithm of an electron can generate force and flux components oriented in more than one direction, assuming a certain complexity. The negative electric charge of the electron probably results from the specific algorithm of this kind of particle. This movement algorithm, in conjunction with algorithms of the same architecture (other electrons), must form a movement away from similar individual cells. Eventually, this movement is balanced by transient particles, produced by the interaction between several types of stable or transient algorithms inside the atom, resulting from the forming flows of typical stable cellular particles such as quarks confined in protons and neutrons, in addition to the interaction caused by the electron flows themselves.

Algorithmic protons and neutrons

Protons can be conceived as algorithms resulting from a specific combination of the algorithms that form the different species of quarks identified by particle physics. In the same way, AMT postulates that neutrons are also algorithms, eventually derived from another form of specific interaction of the algorithms that constitute their specific quarks. Each atomic nucleus is also an algorithm, which combines the algorithms of protons and neutrons, and also the eventual algorithms of the gluons postulated as unstable particles (without isolated existence in nature) that maintain the cohesion of the nucleus.

Algorithm of life

The stability of nuclear algorithms varies according to the chemical element considered. Electronic algorithms are incorporated into the atomic algorithms of different chemical elements. Atomic algorithms involve the loss of electrons and the association with algorithms of other chemical elements, and even with algorithms of the same chemical element, evolving in complexity to form ionic bonds, covalent bonds, and different types of molecules. The evolution of the motion algorithms that form matter crosses the boundaries between physics and chemistry and between chemistry and biology, reaching one of its most advanced stages when reaching the shape of a DNA molecule. In this evolutionary stage, a level very close to the fullness of the traditional concept of algorithm is reached, because a single DNA molecule represents a highly complex algorithm, which, in the right environment, processes the movement of dark matter particles at various levels of structuring. This algorithm has the incredible property of not only constituting its molecular matter but also replicating it, in addition to determining all the tissue specializations that will result in a complex living organism.

Biologists say that in a few years of our life, we renew most of our atoms, basically through food, digestion, cellular reproduction and excretion, among other biological processes. This means that when we look at a photograph of us from about five years ago, that image represents a very different person from the person we are today, at least concerning the particles of matter we are made of, because of mechanisms such as cell death and its renewal. But as a rule, we know that we are the same person as before. This is due to our biological algorithm, which maintains our basic structure, in a process of constant renewal of our matter, despite the effects of the natural aging process.

This is an interesting thing to reflect on. If the AMT is correct, this renewal process is incredibly more intense than we usually imagine. When we think of a single electron in our body as an algorithm for processing the movement of many dark matter particles, which are continually sucked in by this electronic algorithm, and soon after, after going through the circuit determined by it, are abandoned somewhere in the space, being swallowed or not by the algorithm of another particle in our body or by the algorithm of an air molecule, or of some object that surrounds us, it is not difficult to conclude, for example, that the keys of my computer can be exchanging a myriad of dark matter particles with my fingertips as I type these words.

The inertial algorithm

Inertia is one of the greatest mysteries of the universe. We know it exists, and we know its effects, ever since Galileo perceived it as one of the most important properties of the physical world, but we still haven't discussed why it exists or how it operates. We simply accept it as a fundamental feature of the universe we inhabit. Algorithmic Theory also does not have a definitive explanation of its exact mechanism, just as it cannot offer a definitive explanation for the apparent constancy (at least at the local level) of the speed of light, and for many other regularities. At first, the tendency towards constancy of magnitude and direction of velocity, both of photons and material bodies free from the action of external forces, could be seen as a mere whim of nature, something that I consider remarkable. But this can also be seen as something not so strange after all, in the same way that we accept that the velocity of a vase at rest on the table remains unchanged. In general, we can accept that the regularities of the universe result from a natural tendency towards balance. The state of motion of a body, the speed of light (which we postulate, as one of the possible scenarios, tends to constancy only from a local point of view), mass, electric charge, specific heat, hardness, index of refraction, the material liquefaction temperature at a given pressure level, and an infinity of other properties of the physical world that tend to regularity, if considered in isolation, can hardly have a satisfactory specific causal explanation in a short-term horizon. The reasons why certain regularities occur and others do not, and the explanation for the typical values of certain variables that allowed the emergence of life and consciousness can only be explained, perhaps, when we have a broad knowledge of the main dynamics of nature, associated at every stage of the evolution of the cosmos as we know it. Why doesn't a vase on the table suddenly fly away, swell, and liquefy, under mild conditions? What is the causal explanation for all this *not* happening? The most attractive answer is that its state of motion, its volume, and its physical state are in balance with the environmental conditions that surround it. The next big step for science may be to be able to explain this balance at a deeper level, the microscopic mechanism involved in the dynamics that determine it. I believe that the Algorithmic Theory is the key to the description of these dynamics, even if it cannot immediately offer a specific reason why each one of them, in particular, is established. Just one general reason, which since ancient times has been attributed to nature: its tendency to produce balance, although this balance is always destined to be broken by an inevitable succession of events. Balance is the result of imbalance and vice versa. Like Heraclitus and Descartes, I believe that all the physical structures we perceive in the universe are made up of stable flows. This stability, however, is always precarious, as freedom of movement causes the complexity of interactions from which the inevitable imbalance arises. Motion is the fundamental property of all material existence. In our theoretical model, space is not included in this material existence, consisting only of an inert theater with three infinite dimensions where existence is also processed in infinite dimensional scales.

In the case of inertia and the constancy of the speed of light, I consider it appropriate to speak of algorithmic rest, that is, a natural process that produces a particular kind of equilibrium sustained by dynamic interactions, as occurs with the great spot on Jupiter and with the vase on the table. From this perspective, it is possible to imagine that atomic matter at a constant speed and the propagation of light share the nature of algorithmic rest, grounded in underlying dynamic interactions.

By postulating the algorithmic nature of matter, and the intimate relationship between matter and the cosmic medium from which it derives, we realize that from this model naturally emerges a physical explanation for the inertial mechanism.

After all, inertia refers to the state of motion of a material entity in the cosmos. If the cosmos is occupied by an algorithmic ether that produces matter from its movement, maintaining the state of motion of matter in the cosmos presupposes a state of dynamic equilibrium between the material object and the ether that forms it. In this case, we have a particularly curious situation. The translative movement of a material body means the combination of at least two situations. A material body moves through the ether at the same time that the ether moves through the body, giving it physical structure. Ether particles flow not only through the spaces between the material particles of the body but also through the algorithmic circuits that bring these particles into being. Galileo's law of inertia tells us that a material object maintains its state of motion, read modulus and direction of its velocity if it is not subjected to the action of some force capable of changing this motion.

In the Solar System and probably in any other part of the universe that we know, it is almost impossible to conceive of the absolute absence of external forces acting on a material object. But the next question is: how can the Algorithmic Theory justify the *tendency* of a material body to proceed in a straight line in space at the same speed at which it was found when the action of the forces that gave it the last considered impulse ceased? The answer is that a material algorithm is closely linked to an inertial algorithm when the body is in translational motion in space. Somehow, nature determines that an inertial algorithm is in charge of keeping the direction and velocity of the material body stable, in a kind of algorithmic rest, here conceived as a flow equilibrium, until new forces determine a significant change in its state of motion.

Two rocky bodies collide in space, very far from the Solar System's domain, in a relatively empty region of the Milky Way, and a fragment of one of them is thrown in any direction. According to the Algorithmic Theory, we can deduce that the material algorithm of this new rocky body, composed of the various microscopic algorithms of its particles, attracts the sustaining matter as the object crosses space. The movement of these dark matter particles is channeled, within the body, through complex interactions between the numerous material algorithms of its particles, and between these and the cosmic medium that permeates the body, in order to maintain the material structure of the rocky object, its physical cohesion, and its integrity. This orderly internal movement probably coexists with the chaotic by-products of all this movement. So far, we are talking only about the material algorithm. It is to be expected, however, that the greater the velocity of the body in relation to the particles of the cosmic environment that are close to it, the movement of the body causes a more intense effect on the movement of the particles of that medium, in the external region of the body. Not all the movement of the support particles is directed exclusively to the micro-algorithms responsible for the material structuring of the body. As the body approaches a portion of particles of sustaining matter, the effects of this approximation are already felt at a certain distance, altering the movement of these ethereal particles. Just like the winds that announce the arrival of a typhoon, these dark matter particles are set in motion, generating a specific flow caused by the approaching body, always remembering that a part of the ether particles affected by the object's approach can be induced to form an ordered flow while another part can remain in motion chaotic. The dark matter particles taken by this approximation flow will not necessarily be taken to go through the material algorithms of the errant object or cross its interstices. It is perfectly conceivable that this approach flow is subdivided into two other flows, a peripheral one, which passes off the rocky fragment and an internal flow, of a material nature, which integrates the material body. This external flow, caused by the displacement of the body, can contribute to the inertia of the object, feeding its translational movement while also feeding on it. The image to remember here is that of a wind tunnel. In this perspective, a specific additional flow component is formed by the object's movement, and this component, in turn, helps to keep the object in motion. This image is relatively simple, but the real picture can be a little more complex if we consider that the flow component external to the body, if significant, can be responsible for a minimal contribution to the inertia of its movement. Let us remember that in the model in question, there is an external component of flux relative to each particle of atomic matter of the material body. Individually, a particle of support can be found either on the periphery or inside a particle of atomic matter, or even beyond the limits of the body. The most important thing is to verify that in this scenario all the micro components of the flow of material particles inside the body, associated with a possible macro component inside and around the body, together respond to the inertial phenomenon. The feedback of the state of movement results, in this model, from a set of algorithmic reactions of the material body to the effects that its displacement causes in the sustaining medium. It is not surprising that the greater the mass and velocity of the body, the greater the amount of energy required to change the momentum, since the greater the energy that the movement of the body will transmit to the cosmic environment, mobilizing it to maintain the inertial flow. A sustainable algorithmic mechanism of the speed of translational motion of the body is created, that is, a sustaining of the state of movement induced by the movement itself. The inertial algorithm, in this scheme, is in charge of conserving the kinetic energy of the body in displacement, in algorithmic equilibrium with the kinetic energy of the system formed by the body and the medium mobilized by it. This may be the mechanism underlying the Einsteinian idea that an increase in the speed of a body causes an increase in its mass, since nothing, in our perspective, prevents the energy mobilized by each material particle of the body from varying as a function of the body's speed.

To illustrate the idea, we can resort to a very grounded example. We can think of a helium gas balloon, lighter than air, the kind that children hold by a string in amusement parks. In the city of Rio de Janeiro, we use simply call them gas balloons. If a child accidentally lets go of the string, we will see the colorful balloon rising. If there is no wind, the balloon will rise almost in a straight line, at a practically constant speed, I imagine. We know that even though the gas in the balloon is lighter than air, it has weight and is subject to the effects of Earth's gravity. So why does the balloon rise? This is because air is denser, and heavier than helium, and its molecules are not willing to allow the balloon to remain below them. They claim the right to remain closer to the surface of our planet because they have a greater mass per unit volume, therefore, a greater specific weight. But if the air is heavier than the balloon, how does it manage to push its way upwards instead of being stuck to the ground, compressed by atmospheric pressure? We can conclude that some algorithmic mechanism is the answer, associated with the permeability of a particulate medium like air. We can suppose, without being specialists in the subject, that the air molecules in contact with the upper part of the balloon, suffering the effects of the Earth's gravity with more force than what happens with the gas molecules of the balloon, move on the surface of it, looking for a way down. This could put pressure on the air molecules around the balloon, compressing them a little more than normal, and a part of this pressure would be transmitted to the air molecules that are in direct contact with the bottom of the balloon. We would have a situation where the air molecules above the balloon would move apart, compelled by the Earth's gravity, and the air molecules below the balloon would become more concentrated, which would cause a decrease in air pressure at the top and an increase in pressure of the air below the balloon, pushing it upward. Maybe things don't work out exactly as I've just described. This description was not based on a thorough understanding of the dynamics involved in moving a gas balloon through the atmosphere. Possibly the initial part of the description is not necessary to show the difference in pressure that makes the balloon rise since we know that even without the balloon, atmospheric air pressure is greater at lower altitudes. Therefore, the air pressure in the upper part of the balloon is already naturally lower than in the lower part. If we take the example of liquid water, the molecules of which are known not to be compressed easily, we will certainly not be able to use the same imaginary algorithm that we use for the air balloon, even if we know that a full soccer ball, which manages to break free from the cabin of a newly sunken boat will have enough force to displace water molecules that are on their way to the surface of the ocean, as well as air bubbles that also manage to break free from the submerged boat. The most probable cause of the upward displacement both in the case of the gas balloon in the atmosphere and in the case of the air bubbles in the water is perhaps simply the greater pressure that exists in the lower parts of both considered media, causing an upward buoyant force greater than the cohesion forces between the molecules of the immersion fluid. Even so, there are algorithms for the movement of molecules in the medium in both cases.

Be that as it may, in both situations there is a downward flow of molecules from the medium, concerning the periphery of the object. At the same time that the object displaces these molecules, opening a path between them, in an upward direction, the molecules in the medium immediately occupy the space that the object ceases to occupy, helping to displace it upwards. It is not only the object that moves to occupy the place previously occupied by the medium, the medium also moves to occupy the place left by the object. The Algorithmic Theory proposes that a pattern of movement comparable to this one is responsible for the inertia of a body in translational displacement in space, in any direction, with due regard for the specificities of the flow configurations involved. The feedback movement pattern of the algorithmic ether, caused by the state of motion of a material object moving in space, we are calling it the *inertial algorithm*. At first, the image of the gas balloon seemed to make more sense when compared with the movement of a body through the algorithmic ether, even considering the interpenetration between the body and the ether and the micro-algorithms of the atomic particles. After insistent reflections, I came to consider more important for the inertial algorithm the sum of the microscopic effects of movement, through the ether, that is, the effect of the material particles of the body, algorithmically linked together to form the structure of the corporeal object. We thus begin to propose that, in the inertial phenomenon, a deeper integration between the material algorithmic structure of the body, at the microscopic level, and the effects that the translative movement of this structure causes in the ethereal medium as it feeds on it. I applied to the initial model, which assumed an important macro-flow around the body, questions similar to those I offered in relation to the macro-deformation of the space-time tissue postulated by General Relativity.

The material component appears to be of crucial importance in maintaining the state of motion. The inertial algorithm is obviously not exactly determined yet, we are just beginning to speculate about how it can be conceived, like the various material algorithms. However, in the algorithmic model, the existence of a close relationship between the inertial algorithm and the material algorithm seems a natural circumstance, as both algorithms mobilize the sustaining medium particles, constituting an integrated and unique mechanism.

From this point of view, if we consider that all material bodies have some translational motion through cosmic space, we must conclude that the material algorithm always has an associated inertial component. Even in the accelerated movement of a body, the algorithmic premise requires an inertial component associated with its material algorithm, since we are always talking about a movement through the ether of sustaining matter, except for the hypothesis that we consider unlikely, but not completely disposable, of a body at perfect rest relative to an ether flowing into it, at some time and under some circumstances, evenly and smoothly from all directions. We need to keep in mind that in a universe full of celestial bodies like planets and stars, black holes, galaxies, and their clusters, accelerated motion is the most common. Even if a body eventually maintains a rectilinear and uniform movement to a system of coordinates originating in any other object, it may be in accelerated movement concerning many other systems of coordinates, fixed in objects of great mass, even if at great distances. In our planetary system, for example, the orbit of the planets is elliptical. This means that the motion of the planets is accelerated, in magnitude, when they approach the Sun and decelerated, the same, as they move away from it. Even so, we realize that inertia is present in the movement of these planets. Incidentally, it is inertia, combined with macro and microscopic material effects, that allows the planets to maintain their orbits indefinitely in equilibrium around the Sun, preventing the enormous gravity of our star from invariably causing the planets to collide against it, in which case there would be no planets in a stable orbit around a star. Given this, that is, the premise of the algorithmic nature of matter and the fact that inertia is a property of nature attributable to any state of motion of a material body, even accelerated relative to a massive object, we see no reason to dissociate the material algorithm of the inertial algorithm, both closely linked to each other.

The Algorithmic Time

Physics does not yet have an uncontroversial hypothesis about the nature of time. From the mathematical speculations of Poincaré, Fitzgerald and Lorentz, to the famous and firm position of Einstein, time ceased to be absolute, as conceived by Newton, to become a relative concept, varying according to the conditions of the portion of the universe considered. This portion of the universe can be a region of the space-time fabric itself, in the relativist conception, in the vicinity of a material body of great mass. But it could also be the region of space-time occupied by a small material object traveling close to the speed of light. In both cases, the portion of the universe cannot be exactly static, it has to be in displacement, either due to the movement of the massive body, in the first case, or because of the displacement of the modest material body, in the second. In both cases, the bodies carry the interstices of their material substances with them.

Relativity states that the pace of time slows down in the vicinity of massive bodies like the Earth and the Sun. At low speeds, the closer to a planet, the slower an ordinary clock will be, regardless of its model and accuracy. Similarly, the greater the speed of a material object, in relatively deserted regions of large celestial bodies, the slower a clock will count the seconds, until time almost stops at a speed very close to that of the light, relative to an inertial frame of reference, assuming that a simple object does not disintegrate due to its extreme velocity.

Of course, to a human wearing a watch, at least within certain speed limits I suppose, the rhythm of time will appear normal. Just looking at his watch, he will not have the perception that he will be traveling at close to the speed of light (supposedly). From his point of view, according to Special Relativity, the light that appears on the dial of his watch, depending on the position considered, moves away or approaches at the speed of light *(c)*, concerning his eyes. The fact that its speed does not differ much from the speed of light can only be noticed by another observer. According to Special Relativity, an apparent paradox would occur. Let's call the first observer Bia and the second Leo and assume that both move, one relative to the other, at a constant speed, in a region of cosmic space hypothetically empty of other bodies of matter. To Leo, Bia is traveling at close to the speed of light, and her clock would seem slower than his if he could somehow compare both dials effectively. For Bia, Leo is also moving at a speed close to the speed of light, and she would realize (if she could) that it is his watch that keeps time slower. Special Relativity doesn't favor either Bia's or Leo's point of view, so which of the two clocks ticks time more slowly? Would this depend on the pace of time for other observers who would present themselves as judges of the issue? Different observers, moving at different speeds, could they disagree about which of the two clocks is slower and about what is the proportion of the difference in pace between them? For Special Relativity, time is not absolute and there cannot be a privileged observer, subject to a temporal rhythm that can serve as a universal parameter of comparison. The only universal parameter would be the speed of light, constant for any inertial reference considered.

In our algorithmic view, time is also relative. The first variable to determine the variation of the pace of time can be the dimensional scale considered. As we have already seen, we postulate that for each dimensional scale, there are particles constituted by the movement of other particles, belonging to a lower dimensional scale, in an infinite size regression. To understand the concept of time, in the algorithmic view, we must first define in which dimensional scale we are working.

Let's consider our dimensional scale. If for a group of biological beings like us, the rhythm of time becomes slower, this means that our metabolism will process more slowly. Our molecules will interact with each other more slowly, and even the rhythms of translational motion and hypothetical rotation of our electrons will become less intense. All of our matter particles will be moving more slowly. All relevant algorithmic flows on the atomic scale of our corporeal particles will have lower velocities. The same will happen with the particles of the clocks attached to our wrists. That's it, in the end, that will make its pointers or digital dials slow down. In this context, time can be defined as the rhythm of movement patterns. This means that time is also algorithmic.

The concept of time, for us, as well as for different philosophical currents throughout history, presupposes the existence of cyclical and regular movements. Without something moving in comparable repetitive patterns, one cannot conceive of time-evolving counting units. If we speed up or slow down these cycles in a generalized way for a portion of the universe about some frame of reference, we can speed up or slow down time in the observed region.

Let us call first-order dark matter the particles of the algorithmic ether of the scale immediately below the scale considered. If we could reduce our size down to the size of a single first-order dark matter particle – and if even then there were some kind of light visible to us – we would see, on the algorithmic hypothesis, that this particle is also formed, in a simple case, by a swirling flow of much smaller particles. Comparing the hypothetical rate of rotation of an electron of our human scale – without reducing our size – with the rate of rotation of a first-order dark matter particle, we can easily imagine that this smaller-scale particle makes several turns around its axis while the same particle travels a single translational turn in the electronic vortex of our scale that it eventually participates in, assuming that the translational distance of that turn is brutally greater than the diameter attributed to the dark matter particle. This comparison is certainly somewhat arbitrary, but it can be obtained by the inductive method when considering various movements that we observe in nature. As an example, we can compare how many times the Earth rotates on its axis while the entire solar system takes a single turn around the Milky Way. This example helps us to visualize time as it has almost always been done throughout history, comparing cyclic patterns of movement, and determining the relationship between them, by defining units of comparison, such as the day and the earth year.

Let us now imagine that all the particles of normal matter that we know disappear. We are basically talking about electrons and quarks, and even force-transmitting particles, such as photons and gluons. According to the Algorithmic Theory, at least until these particles return to form, time would cease to exist in the dimensional scale in which we live, but this would not mean the absolute non-existence of time, as it would continue to exist, in the algorithmic hypothesis, at least on lower dimensional scales. The destiny of the superior scales would accompany that of our scale, due to the inexistence of particles that could form the particles of the superior scales, or perhaps those scales could continue to exist if formed by particles of scale similar to ours, found in neighboring portions of the universe, eventually not annihilated by the disintegration of electrons and quarks that we know. As a result of the fact that motion cycles on the scale immediately below ours – the hypothetical rotational motion cycles of dark matter particles, for example – are faster, due to the smaller spatial dimensions involved, we can say with a good deal of security that the time would continue to elapse in lower dimensional scales than ours, but at a much faster pace, based on much faster cycles, since they are much smaller. In this sense, it is not absurd to speak of multiple temporal dimensions, as long as they are associated with the dimensional scales proposed by AMT.

Leaving aside the temporal variation between different dimensional scales, our Algorithmic Theory states that if something is capable of altering the rhythm of the algorithms that it claims as characteristic of material objects, it will automatically be capable of altering the rhythm of time for those objects as well.

Algorithmic acceleration

Relativity proposes that two variables are capable of modifying the pace of time. The velocity of the object and the gravitational intensity to which it is subjected. Increasing its values produces an effect of decreasing the pace of time. One of Einstein's most extraordinary findings was that the effects of gravity are equivalent to the effects of the acceleration of movement. But what would be the physical ingredient capable of making these effects equivalent? What would the gravity have, physically, in common with the acceleration of movement? Relativity does not seem to offer a strictly mechanical satisfactory response, although its mathematical response has great relevance to science in the historical and quantitative aspects. The lack of a genuinely physical conception, in the most classic sense, that explained the success of the applicability of mathematical results, certainly contributed to the adoption, by Einstein, of a geometric conception to the very physical nature of the universe. The Algorithmic Theory aims to be a conceptual instrument capable of offering a realistic physical explanation for the nature of the universe, which can be described by mathematical models that represent it, at least for a considerable range of dimensional scales (three or four would be an extraordinary achievement).

The premise of the existence of algorithmic ether, stratified in successive dimensional scales, seems to assume naturally the physical explanation for the equivalence between gravity and acceleration of movement, perceived by Einstein. But it can also serve as an explanation for a scenario where there is no physical equivalence and that only the effects of gravity and acceleration of movement are similar under the specific conditions of Einstein's famous mental experiment. As we propose, the acceleration of a body or particle of matter represents nothing more than interconnected or individual material algorithms in the course of a change in its state of equilibrium with the sustaining medium. If an external agent induces acceleration or deceleration for some time, the material algorithm alters its energy interaction in the form of kinetic interaction, flux interaction, with the algorithmic environment. By the time the agent's action, the material body or particle is already adapted to the new energy and inertial condition of its material algorithm, showing a new translative speed in a new balance with the support medium. This new balance will be characterized by greater or lesser intensity and range of the mobilization that the material algorithmic tissue causes in the medium of which it feeds.

Algorithmic strength

Through this concept of acceleration, the force can be defined as the action of any agent that manages to alter or sustain (against the action of another force), algorithmically, the inertial rest of a material object. Once the limit of structural resistance of the material body is exceeded – the limit of its ability to adapt to the effort to which it is submitted – the force causes the rupture of the composite algorithm responsible for the integrity of the object.

Acceleration and gravity

The equivalence that Einstein realized refers to the comparison between two closed laboratories (or elevators), one at rest, subject to a gravitational field and one in an environment without gravity, but subjected to constant artificial acceleration equivalent to the acceleration of the gravitational field considered. Anyone who is on each laboratory will have the feeling of being firm on the floor. If objects are dropped at a certain height of the floor, in each laboratory, they will fall free with the same acceleration. So far, there seems to be nothing to discuss. I believe, however, that equivalence resides only on the effects of the experiment. I have for myself that we are dealing with two completely different phenomena. It is one thing to drop a tennis ball in a laboratory on the earth's surface. The ball falls because of an active effect of gravity. Another thing is to drop the tennis ball in an accelerated laboratory in space where there is no gravity. In this case the ball falls because it stops accelerating and it is the laboratory floor that approaches the ball, as it remains accelerated. I think the most interesting thing about this is trying to understand why gravity makes the tennis ball fall.

Gravity is comparable to an increase in "temperature", an increase in chaotic kinetic energy, caused by a material object in the support ether inside and around the body, in an energy gradient that decreases with the distance from the center of mass of the material body.

Material or energetic algorithms (such as light particles) have an attractive affinity for the region of higher chaotic kinetic energy of sustaining ether. An algorithmic affinity. It is as if matter and energy were attracted to the "hotter" regions of the cosmos (denser regions in chaotic kinetic energy). In our theoretical model, denser regions of chaotic kinetic energy, in the algorithmic medium, besides attracting the material algorithms, are also responsible for the slowdown of time in these regions. It turns out that in the presence of more energetic ether particles it becomes more difficult for material algorithms to attract ether particles to the fluxes of matter or energy particles. Capture becomes more costly and the whole system gets slower, and this is the factor responsible for the slower pace of time in regions where gravity is greater. The metaphor is to try to stir a very thick oatmeal with a wooden spoon. The same effect can occur with increasing speed. When a body moves faster compared to ether particles, it is as if the sustaining particles are more stirred over the body, making it more difficult to capture by material algorithms.

The apparent paradox of algorithmic gravity

We said above that the material algorithms of the bodies are attracted by the densest regions in terms of chaotic kinetic energy of the algorithmic medium. We also said that when penetrating the denser regions in chaotic energy of the medium, these same algorithms become slower, which causes a slowdown in the rhythm of time. The two statements taken together may seem to form a paradox. If greater kinetic density attracts algorithms, how can this same density make these same algorithms more difficult and slower? I believe there is no problem in admitting the two hypotheses together. We are again faced with a metaphysical question. Both things need to happen in nature for AMT to be consistent and compatible with the relativity of time. It is a matter of belief, of intuition. Only mathematical models and computational simulations will be able to corroborate or not the theoretical proposal as it is presented. But before such models and simulations are available, I once again make use of the image of the slipper dragging a plastic ball on the surface of the water, mentioned earlier. As the ball is dragged along by the effect of the slipper, attracted by it, I fancy that it would be more difficult to do the work of removing the ball from the water than if the ball were in a calm region of the surface. It is the best example I can give at the moment to illustrate the point.

Algorithmic relativity

This model has a consequence that challenges a fundamental premise of the Theory of Relativity. Although compatible with the relativistic model regarding the factors that cause the deceleration of time (gravity and velocity), the algorithmic scheme contradicts the view of Special Relativity that the difference in the rhythm of time would be symmetrically opposite for two bodies with displacement between them, at a constant speed, even in environments with sufficiently negligible gravity, that is, without considerable gravitational influence from other masses in the region where the hypothetical experiment takes place. The apparent paradox of Léo and Bia's clocks referred to in the section "Algorithmic time", above, does not occur according to our theory. For us, the change in the rhythm of time does not result especially from the relative movement between two objects, and does not necessarily occur at the same intensity for two objects that maintain a constant speed of displacement between them. It is necessary to consider the speed of displacement in the local ether for each body. It is necessary to consider this *privileged local referential*, not admitted by classical relativity, which behaves dynamically, *the privileged referential formed by the medium the flow of which sustains the matter of each object considered*. Furthermore, it is the individual relationship of each object with the algorithmic medium that surrounds it that will determine the temporal rhythm in local conditions. Einstein held to the relativity, individuality, and locality of time. Our theory accepts such attributes of time and adds that they *result from the dynamic conditions of interaction between the particles of a material individual and the particles of the ethereal medium that sustains the materiality of this individual and that reacts to the intensity (speed) of its displacement in that medium*. In our theoretical framework, we are forced to discard the paradoxical symmetry of changing the rhythm of time that Special Relativity associates with uniform motion.

If we sent Leo and Bia in the same rectilinear trajectory starting from a point of space and left observers at points in the trajectory, we would see that Leo's clock would record the evolution of time more slowly than Bia's, if Leo moved at a speed bigger than Bia's, even if speed of each one remained constant. The delay of Leo's clock in relation to the clock of each static observer would be proportionally larger than the delay of Bia's clock. These results would derive from the already discussed premise that the greater the intensity of the object's algorithmic performance over the algorithmic ether and vice versa, the lower the rhythm of the material cycles that determine the units of time for this object. We can make an analogy with two boats moving on a river, piloted by two boatmen. The one that moves faster relative the water molecules (Leo), regardless of the direction of the current and the speed of the flow of the water about the banks, would have its clock marking time more slowly than another (Bia) that turns off the engine and lets the boat go with the current alone, moving away from the other boat at a constant speed. It should be noted that this analogy is full of obvious and natural imperfections, used only as an illustrative resource.

Kinetic concentration (or kinetic density) of the algorithmic medium

The ordered (algorithmic) movement of the particles of a massive object causes a large chaotic movement of the first-order particles in the sustaining medium. There is a great concentration of both chaotic and ordered movement in the space occupied by the body. Chaotic motion is also induced beyond the limits of the material body. Naturally, the closer to the center of the body, the greater the kinetic concentration of particles of the ethereal medium. A decreasing, chaotic kinetic concentration gradient is formed as we move away from the center of the body. This kinetic concentration gradient can also be called gravity.

At this point in the narrative, we have already abandoned the initial idea of a possible destructive consumption of dark matter particles by massive bodies, as well as the hypothesis that a smooth flow could be responsible for the gravitational effect. The idea that the chaotic kinetic concentration caused by material algorithms in the interstices of a massive body is responsible for the body's gravity seems largely advantageous to us, also because it allows dispensing with the idea of destructive consumption of the sustaining ether. The idea that a bombardment of smoothly flowing ether particles would cause gravity no longer serves us. In its translational movement, a massive body mobilizes order and chaos in the sustaining ether in all directions from the center of the body, taking with it the chaotic mobilization that it causes in its surroundings and its interior, thus carrying its gravity wherever it moves.

On the other hand, we return to the idea that the internal movement of bodies with mass attracts other bodies with mass, with the difference that we are no longer considering the movement of molecular and atomic agitation as an attraction factor, but the chaotic movement of the sustaining particles present in the interstices of the particles of matter that form the body and the ether agitation that goes beyond the very limits of the body in question.

Algorithmic constancy of the speed of light

The Algorithmic Theory gives rise to the idea that the constancy of the speed of light is probably not absolute and can be influenced by gravitational conditions. Like all matter and energy, why wouldn't light be subject to the gravitational influence, including concerning the constancy of its speed? I believe that the constancy of the speed of light can manifest itself only under local gravitational conditions. It implies that along its path through space, its velocity may vary, as there are no uniform gravitational conditions over great distances. This does not prevent gravitational conditions from being more or less uniform on the surface of a planet, as in the case of Earth, where the interferometer was used in the famous experiment by Michelson and Morley.

If this idea is correct, we envision at least three possible scenarios (hypotheses) for the behavior of the speed of light as a function of local gravitational conditions.

First scenario

The first hypothesis is that in the presence of strong gravity, light accelerates its movement when its trajectory is directed toward the gravitational center. This is the behavior of material bodies, which accelerate when in free fall, without ceasing, when there is no atmospheric resistance. We know, as Einstein proved, that light is attracted to matter and can fall into large objects like a black hole. The question is whether, under these conditions, light can also accelerate, like matter. If a particle of light is a movement algorithm, as we propose, and if the other particles of movement accelerate when falling, it is not strange to suppose that the same happens with light, but with acceleration presenting greater difficulty of detection by human technology than occurs with corporeal objects like an apple.

Second scenario

Corporeal objects accelerate in free fall, but light slows down. In this case, what would justify the behavior opposite to that of material bodies? The answer may lie in the hypothetical fact that, being the speed of light the limit of translational displacement, per unit of time, that anything can reach, as proposed by Einstein's relativity, the speed of light, when decelerating in free fall, could describe a curve of maximum values of velocity that matter or energy can reach for each specific gravitational intensity. The maximum values of velocity for each value of gravitational intensity, in this hypothesis, would be decreasing, as gravity increases, which would slow down the light. In this model, bodies in free fall continue to accelerate because they cannot reach the speed of light in any instant. Here it is worth noting that in this second scenario, light moving away from a massive body accelerates as the gravitational chaos of the algorithmic ether becomes more rarefied, that is, as gravity decreases.

Third scenario

At least within certain limits of gravitational intensity, light does not change its constant velocity, both from the point of view of the photon and the point of view of an observer in an inertial frame of reference, always remaining at the known constant velocity *(c)*. This constancy would be provided by a compensation mechanism in time and distance measurements, such as that of Lorentz and Fitzgerald, assumed by Einstein.

Other possible scenarios

Alternative scenarios may involve a mixture of the two or three basic scenarios presented above. In this way, we can imagine that the speed of light remains constant or increasing and, from a certain value of gravitational intensity, begins to decrease, in addition to other possible variations of combinations of the first three scenarios.

The important thing is that the Algorithmic Theory of Matter opens up several possibilities for the behavior of the speed of light as a function of gravitational intensity. Only mathematical and computational models, in addition to eventual experiments with fluids, can support a model that becomes acceptable to indicate what in fact may occur in nature.

CHAPTER V

SOME DISCOVERIES ALONG THE WAY

Some discoveries along the way

Regarding the development of AMT, I want to reveal some discoveries that occurred during the studies concomitant with the elaboration of this book. While I devoted myself, in my free moments, to writing the first few pages, my desire to read everything related to the frontiers of modern physics increased. In the first lines, the feeling prevailed that no one had dared to challenge the non-existence of the ether since Albert Einstein tried – and practically succeeded – to promote its official banishment from physics. I discovered, however, that the debate about the existence of some kind of ether never completely died out. From a personal perspective, after having consolidated, through previous readings, the impression that this challenge would be an isolated attitude that I should assume, I got a huge fright when I found an old issue of a well-known publication in a bookstore, which shook my belief as to the originality of the Algorithmic Matter Theory.

The work of Jacobson and Parentani

In the middle of the first half of 2012, maybe in March or April, in the center of Curitiba, I passed by a used book and magazine store. On the shelf closest to the entrance, there were several issues of Scientific American Brazil, one of them highlighting, on the cover, an article on dark matter. I went into the bookstore and started to separate all the copies with cover subjects that brought discussions about space-time, subatomic particles, and related topics. One of the copies was a little damaged, and I asked the attendant if there could be another one in better condition. I remember the structure of the store well, the position of the magazine shelves at the entrance, on the right, and the service counter, more in the center of the establishment, in front of the shelves. The kind employee directed me to the back, where there were still several piles of old copies of that same periodical. I found a more complete copy of the issue I was looking for, but I couldn't help but take a good look at all the copies piled up in that section. Behold, among them, came one that caught my attention. The title of the main article, which accompanied the cover design, was "Echoes of Black Holes". In the center of a dark circle, representing a black hole, was the subtitle: "Sound waves give clues that space-time may be a type of fluid". At this point, the reader friend can imagine how much my heart accelerated. It was lunchtime, on a working day, and I started devouring right there the text of the extensive cover story, signed by Theodore A. Jacobson and Renaud Parentani. The issue had been published in Brazil in January 2006. On page 43, the internal title was "Propagation in Black Holes", accompanied by the following introduction: *"Sound waves in a fluid behave remarkably similar to light waves in space. There are even acoustic equivalents of black holes. Now the question is whether space-time can be a kind of fluid, like the ether of pre-Einsteinian physics."*

I had been searching for several months and had never found anything like that. Official physics considered the rebirth of the pre-Einsteinian ether. The article struck me as absolutely remarkable for apparently suggesting answers virtually identical to those I had been formulating for a series of questions. Note the first three paragraphs of the article:

> *When Albert Einstein proposed the special theory of relativity in 1905, he rejected the idea, common in the 19th century, that light resulted from vibrations in a hypothetical medium called "ether". Instead, he argued, light waves can travel in a vacuum unsupported by any material – unlike sound waves, which are vibrations in the medium in which they propagate. This particular of general relativity [17] is not touched in the two other pillars of modern physics, general relativity and quantum mechanics. So far, all experimental data, from sub nuclear to galactic scales, are successfully explained by these three theories.*
>
> *However, physicists face a profound conceptual problem. As understood today, General Relativity and Quantum Mechanics are incompatible. Gravity, which General Relativity attributes to the curvature of the*

---

[17] Read *"This particular of special relativity"*.

I started writing this section on Friday night, 6/14/2013, and I am re-reading Jacobson and Parentani's article. I feel the same butterflies in my stomach that I felt that afternoon in March or April 2012, even more so because I have just noticed details that increase the proximity of the article's conclusions relative to Algorithmic Theory. Furthermore, I had not realized so fully how the article addresses the possibility of varying the speed of light.

But what I read in such a hurry and was able to apprehend that afternoon over a year ago was more than enough to realize the dimension of that work and its almost total compatibility with Algorithmic Theory. Far from making me excited, that discovery caused me an undisguised feeling of frustration, because at first glance, still under the impact of fright, it seemed that those few pages could exhaust the algorithmic proposal. Upon returning to the room where I worked, I showed the magazine and revealed the discovery to colleagues, who had always been

---

[18] JACOBSON, PARENTANI, 2006, p. 43

denied any clue about the theory they knew I was developing. In a tone of disappointment, I announced that someone had already published, six years before, what I was keeping under lock and key, even hiding it from my parents, brothers, and brothers-in-law. Sympathetic to my dismay, some even congratulated me on the fact that, despite not being a physicist, I managed to formulate something revolutionary, very similar to what had been suggested by those world-renowned professional scientists. I now note that the article by Jacobson and Parentani was originally published in May 2002, in Pour La Science, the French edition of Scientific American.

I explained to my co-workers that my theory was based on the existence of an ether of particles, completely ruled out by Einstein, and I read to them the final part of the article, perhaps the following excerpt (emphasis added):

> *Applied to real black holes, the analogy with fluids indicates that Hawking's result is right, despite the simplifications he made. It further suggests that the infinite redshift in the event horizon of a black hole should likewise be avoided due to the scattering of short-wavelength light. But there is a caveat. The Theory of Relativity asserts that light is not dispersed in a vacuum. The wavelength of a photon looks different to different observers; is arbitrarily long when viewed from a frame of reference moving at a speed close enough to the speed of light. Therefore, the laws of physics cannot command a fixed dispersion of short wavelength, in which the dispersion ratio changes from type I to type II or III. Each observer would perceive a different dispersion.*

> *Physicists then face a dilemma. Either they retain Einstein's injunction against a preferred frame of reference and swallow infinite redshift, or they admit that photons are not infinitely redshifted and are forced to introduce a preferred frame of reference. Would such a system surely violate relativity? Nobody knows. Perhaps the preferred reference is a local effect arising only in the*

vicinity of black hole horizons – in which case relativity continues to apply in general.

*On the other hand, a frame of reference may exist everywhere, not just in the vicinity of a black hole - in which case relativity is **just an approximation of a deeper theory of Nature**. Experimental physicists have yet to find such a system, and this may not have happened yet due to insufficient precision in the experiments.*

*It has long been suspected that reconciling relativity and quantum mechanics would involve a shortcut, probably related to the Planck scale. The acoustic analogy reinforces this suspicion. **Space-time must somehow be granular to tame the dubious infinite redshift**.*

*If so, the analogy between sound and light propagation would be even better than Unruh believed. The unification of General Relativity with Quantum Mechanics could lead to the end of the idealization of continuous space and time, and the discovery of space-time "atoms". Einstein may have thought the same way when he wrote to his close friend Michele Besso in 1954, a year before his death: "**I consider it perfectly possible that physics is not based on the concept of field, that is, on continuous structures**".*

*That would shake the foundations of physics, and **today scientists do not have the right candidate to replace them**. In fact, Einstein's next sentence was: "**Then, nothing remains of my castle in the air, including the theory of gravitation; but also none of the rest of physics**". Fifty years later, the castle remains intact despite its uncertain future. Black holes and their acoustic analogues may have begun to shed light on the issue.*[19]

---

[19] JACOBSON, PARENTANI, 2006, p. 49

As I began to recover from the impact, I remembered something I used to say to my beloved wife Sonia while watching episodes of the Discovery Channel series "Through the Wormhole", with Morgan Freeman. I used to say that the more I learn, I realize the less Algorithmic Theory has to reveal, as something original, but truer it proves to be.

What did we have then? Jacobson and Parentani had boldly challenged the space-time continuum of General Relativity, transforming it into an atomized fluid where light propagates with variable speed under certain conditions. From the point of view of Algorithmic Theory, this is an extraordinary historical finding! But how does this fluid work? What is its relationship with matter? How does he explain gravity? How to explain, from this fluid, all the physical phenomena known to man? These questions were not answered by the article. Furthermore, Jacobson and Parentani only suggested, with a high degree of conviction, it should be noted, that "Space-time must somehow be granular...". It remained to explain the mechanism of this "somehow". By pointing out how this can be done broadly and comprehensively, the Algorithmic Theory, to my relief, remained original and intact. The fundamentals of the theory had not been presented in the article: the algorithmic principle of matter and the cosmos, the hypothesis of fluid particles, which can form complex flow structures, the dimensional stratification of the ether, the fundamental role of the algorithmic ether in the essence of phenomena of gravitation, electromagnetism, inertia and all other physical phenomena that we can observe.

I would like to know how far those scientists have gone with their research, speculations and conclusions. They would be some of the first theorists with whom I would want to discuss the algorithmic scheme. Be that as it may, their exceptional work will help accelerate acceptance of the new general scheme to which I am dedicated and which, in part, they were able to anticipate.

I came to wonder why Jacobson and Parentani's work didn't get more publicity, but I soon learned that the sonic black hole analogy is frequently mentioned in television programs and articles on the fundamental topics of physics. In fact, I was the one who had not realized the scope of their research before coming across the words originally published by the authors. And to think that I only found that article because of another magazine with a slightly frayed cover...

Parallel universes? – the bookstore riddle

There was another curious fact about this discovery. For some reason, I don't remember what, perhaps to see if I could find anything new, I returned to the bookstore a day or two later. The store was different. The shelf with the scientific journals was on the left side, next to the wall. The counter was no longer in the center, but at the right of who enters, where the shelf with the magazines used to be. The entire interior of the store was slightly different in appearance. Even the attendant was different. I asked if there was another bookstore of its kind on the same street, as I might have mistaken the store. The employee said that a few blocks ahead there was another bookstore. I walked much further than I thought I had walked the other day and found the indicated store. The structure in no way resembled the store where I had bought the magazines, and the name of the bookstore was also different. I asked if this second store had recently moved there and the girl at the counter replied that they had just moved to that location, but from a location that was not on that street, but at a completely different address. I began to be more and more astonished, as my investigation into what might have happened to the layout of the store was leading nowhere. I went back to the first store and looked at the sign: it was the name that I thought belonged to the store from the other day.

In a last attempt, I asked if they had changed the entire layout of the store that week. The answer was that the store had been like that for several months. If I remember the answer correctly, maybe since the inauguration. I found myself completely helpless. How to explain that paradox? The certainty about the previous form of the store was absolute. Uncertainty about the name of the store was minimal. You know, at these times we try to look for all kinds of explanations. It felt like I had just entered a parallel universe, worthy of an episode of the "Fringe" series, with Anna Torv, where this had just happened to one of the protagonists. That episode was recent, and it was still in my head. The great irony is that Algorithmic Theory does not point to the possibility of the existence of parallel universes. Quite the opposite.

I went back to work trying to carefully observe, in the streets, looking for some small change in the scenery, after all, I needed to test the only hypothesis that occurred to me, that of having been transported to a parallel universe. No visible changes. At work, the people were the same, as were the problems to be solved. No break in continuity. I confess that it was interesting to imagine for a few moments, as in a dream, that more evolved beings were trying to send me some kind of message, a sign that I would be on the way to deciphering the nature of matter and the essence of how our universe works.

I dreamed of something like Star Trek, in the movies, when the Vulcans revealed themselves to the Earthlings as soon as they achieved space folding technology. By the way, this is another point of science fiction that would be incompatible with the Algorithmic Theory, since it is not possible to bend a continuum that does not exist. Perhaps the series of the future will refer to technologies for manipulating the flow of dark matter, and not equipment for torsion of the space-time fabric. The movie "Back to the Future" curiously shows that the device responsible for the success of time travel is called a "flux capacitor", whatever that may have meant in the fiction author's mind.

The magazine store story still needed to be resolved. The evidence of reality pointed to a more mundane explanation, though I didn't know which one.

Scientific hypotheses and lost objects

Perhaps this is a good time to mention a technique I developed for finding lost objects. This technique is nothing original and has probably been used for millennia, but unfortunately, it doesn't seem to be widespread in most people's daily lives. While this book won't revolutionize science and philosophy, I hope it can at least help people find objects like keys, documents, administrative processes, or any other artifact they're looking for. I believe this also applies to scientific hypotheses and lost magazine stores. The rule is:

> ***The object you're looking for is probably a lot closer to where it should be than you think.***

In early 2014, I would find this same rule, referred to as "back to the beginning", in an American film from 2005 or 2006, with the Portuguese title "Caos" ("Chaos"), if I'm not mistaken, in the voice of Matt Damon's character.

The explanation is simple. When we look for something, we usually start from where we think it should be. The place where the object is usually kept or where we remember last seeing or using it. As we do not find it, we expand the search radius and thus move away from the place where he should be. Logic indicates, and my personal experience confirms it, that in the vast majority of cases, the reason for not finding the object is that we have neglected to search the surroundings close to the initial location. In the end, we're looking farther and farther away from where the object actually is. We've been through it and we didn't realize it. To complicate matters, we tend to be victims of a psychological phenomenon that consists of enormous mental resistance to looking again in places where we have already looked. We prefer a thousand times to keep looking in new places than to revisit an already discarded place. I assume that this is related to the great discomfort we usually feel when we are forced to redo any task.

The recommended recipe is also simple. Stop everything if you notice that you have widened the search radius too much. Ask yourself where the object should be. Use logical criteria, memory or even intuition. Ask yourself what clothes you were wearing, check if you kept the object in a pocket, remember the places you went while holding it, but mainly what would be the place where that object should be. Use reason first. If there is no natural place for the object, use intuition. Go to this place and look for it with the certainty that it is there, or at least in some very, very close vicinity. Dig deep, going through the surroundings with a fine-comb. You'll see that you'll almost always find the object in one of the first places you looked for it, but due to haste or carelessness, you didn't check properly. You can try it is an almost one hundred percent guaranteed method. If this book helps to spread this method, I believe that human society will stop allocating many resources to unproductive searches, will avoid considerable losses in replacing lost things and will save many moments of prolonged stress.

At my workplace, which I believe I mentioned was a department that deals with bureaucracy, after I insisted with my closest colleagues that this technique would be useful, some started to adopt it with great results. Once, we carried out an inventory to locate approximately 3,500 physical processes (a form of documentation that is still widely used at the beginning of the 21st century, although it already faces brutal competition from electronic documentation). These physical processes consist of a set of paper documents attached to a cover by two flexible metal clips that pass through the first cover and the set of documents, through two holes. The clips are folded before the second cover, the two covers formed by the halves of a single double sheet, usually slightly thicker.

The processes, after closing the corresponding procedures, were stored by our team in boxes, and sorted by the closing date. Once a teammate and friend, Ana, proudly told me how she found a missing process:

> "I used your technique. I had already looked twice in the box where the file was and started looking for it in all the cabinets. Then I remembered what you said and wondered where the process should be. I went straight to his box and removed all the processes, very carefully, checking if one process could be inside the other or if two processes could be stuck together. Sure enough, the process that was missing was so thin that the outer parts of its clips got caught in another process, and when I pulled the top one with my fingers the bottom one was pulled with it, without my noticing."

There were other situations like this one, in which the technique was successfully applied, and we started to adopt it habitually in our work section.

Solving the bookstore riddle

I don't remember to consciously using this method to solve the magazine store problem, but in the end, that is what happened. After spending a day or two hanging around the "transformed" store, I ended up walking past the store in its "original" form. The name stamped on the sign left no doubt. The point is that the original store was not on the street from the second store, but on a side street, a few steps from where I was persistently looking for it. After recognizing the clerk, the shelves, the counter and everything else with relief, I left the store, turned the corner and there was another store, with the same name and style on the sign, but with another salesperson and a different internal layout.

For those who remember the beginning of this book, can imagine what I am trying to say with these recollections. If some answer needed to be found to unify Relativity and Quantum Mechanics – or replace the explanatory concepts of both models –, nothing better than looking at their common origin, at the beginning of the 20th century, when the ether of particles, still immature, despite being very old, was thrown to the brink of non-existence, where it remained for almost a century to be found again – and who knows, definitely rehabilitated – now wearing a new essence, incorporating in its dynamics the structure of matter, gravity, inertia and other physical phenomena, preserving and justifying what is most precious in Relativity – the relativity of time –, ironically the theory that tried to bury it for all eternity. The irony becomes greater when we remember that without Relativity the algorithmic ether would probably not have achieved the modeling that I am presenting and that I believe will develop continuously and prevail for long ages.

Before mentioning other works that I consider important, with which I had contact in the course of the slow process of producing this book, I consider it fair to mention that I have been mentally maturing some conceptions long before transposing them to paper, or rather, to the computer. Many ideas have already been formulated in thought and have gone through an internal and external dialectic, through readings, before and after starting to write this work.

At the beginning of the Algorithmic Theory registration process, the obsession with ideas made me lose nights of sleep writing, or simply trying to mentally resolve some logical impasses. But, as time goes by, as I don't belong to the gym and I need to dedicate myself with priority to my professional and personal tasks, the compulsion sometimes cools down. I am even afraid that the richness of some ideas will be lost over time or has already disappeared.

Aetherometry

During some phases, I got more carried away with the readings than with the writing. In one of these readings, I was shocked when I learned about the work of the couple Paulo Correa and Alexandra Correa, on some Internet discussion forum. One day, I decided to do a Google search for the word "ether". I ended up discovering that there is what the

Correa call aetherometry, in homage to an ancient form of the word ether (aether). In Portuguese, we can call it Aetherometria, preserving the aether form. From what I understand it is a science that theoretically and experimentally studies the ether, and it originated with Wilhelm Reich and his studies on bioenergy and homeopathy. Correa's work has received harsh criticism, as far as I could see, especially for the insufficient dissemination of their experimental methods, which would make difficult the dialectics about their conclusions and confirmation of their results.

Something I want to highlight about Aetherometry, according to texts attributed to the Correa couple, and which impressed me a lot, were some of their conclusions, remarkably compatible with the algorithmic scheme that I have been developing. Everything is attributed to the ether, including inertia, which surprised me enormously. The explanation of inertia was one of the first logical consequences that I deduced from the algorithmic model and, as I see it, constitutes one of the greatest triumphs of our model, as it allows a detailed microscopic glimpse into how this mysterious and fundamental property of matter can work.

Seeing that Aetherometry also tries to explain inertia from the ether gave me a feeling almost as strong as the one I experienced when reading Jacobson and Parentani's article on the sonic analogy of black holes. And it wasn't just that. Some of the most important conclusions of the Algorithmic Theory were anticipated by Aetherometry, as Descartes had already managed to anticipate, in the sense that ether is the substance responsible for virtually all physical, chemical, and biological phenomena.

Thomas S. Kuhn

In June 2013 I started reading the book The Structure of Scientific Revolutions by Thomas S. Kuhn, although my friend Antonio had already explained to me, before I started writing about Algorithmic Theory, something about the crises that mark the replacement of paradigms and scientific progress, in the Kuhnian view. Antonio was my teammate at work, and he always recommended me good books on the history of physics, since he had attended two years of higher education in this science. Following a peculiar pattern of having contact with books through him, I only started reading Kuhn's book because Antonio took the initiative to lend me his new copy, recently acquired because a cousin did not return the old one he had. It was also through Antonio, whom I thank, that I came across Brian Greene's book, "The Elegant Universe", and the book by Einstein and Infeld, "The Evolution of Physics". I also thank my dear brother-in-law João for having given me, from his private collection, Einstein's book, "How I See the World".

Aetherometry x Algorithmic Matter Theory

I found in Kuhn an explanation of the reasons that kept Aetherometry confined to a few adherents and marginalized concerning the so-called normal science. After all, the ether has been banished from normal physics. But beyond the sociological mechanisms observed by Kuhn, I believe that the discreet prosperity of Aetherometry is due to the explanation that it offers as the foundation of its conclusions. From the superficial and incomplete reading that I did, it seems to me that the mechanisms proposed by Aetherometry are taken from the current paradigms, related to the concept of space-time in its quantum version. From what I could observe, the basic mechanism developed by Aetherometry consists of the frenetic collision between fundamental particles of normal matter that randomly appear in the quantum foam. Although Algorithmic Theory shares much of the conclusions of aetherometry, the algorithmic premise, if true, centered on flow structures, invalidates the mechanism of the aetherometric model by negating and replacing the concept of space-time in both its original relativistic and quantum version, which does not prevent that at every scale of the algorithmic ether there is in constant collision, destruction and creation of the particles of the sustaining medium.

The algorithmic scheme, despite involving different assumptions and mechanisms, does not diminish the importance of the aetherometric model. On the contrary, it elevates it to the category of historical anticipation, albeit a distinct one, towards the model we propose. I find impressive the stubborn intuitive confidence that the creators and developers of aetherometry demonstrated in the existence of an ether that propagates energy, including electromagnetism, and constituent of normal matter, also responsible for its inertia. Seen from this angle, aetherometry becomes a precursor theory of the model we are presenting and, as such, it must be recognized, despite not having influenced the subjective process of creation and development of our model, which, as we have already explained, is directly associated with questions such as the identification of shortcomings in the concept of space-time and the search for a scheme involving dark matter that could explain gravity.

Roberto de Andrade Martins and Alberto Mesquita Filho

In the reconstitution of the discoveries regarding works that I consider important for the theme we are discussing, from readings concomitant with the elaboration of this book, I must also mention two names that helped me immensely to historically contextualize the Algorithmic Matter Theory: Roberto de Andrade Martins and Alberto Mesquita Filho, the latter already mentioned in another topic.

Roberto de Andrade Martins wrote an article in which he claims that *"Contrary to what some people believe, neither Einstein nor any other researcher proved that the ether does not exist. It can never be proved that there is nothing in a certain region of space because there could still be unknown physical beings there, which we do not know how to detect"*.[20]

More from Thomas S. Kuhn

From Thomas S. Kuhn's book, I want to highlight the seventh chapter, entitled "The Response to the Crisis".[21] Reading a book like this, currently, finding myself in possession of the algorithmic model, requires a coldness and a self-control that sometimes I think I am not able to keep, since I would like my own book to be ready, revised and edited, which would perhaps allow me to participate as soon as possible in the great debate that is approaching. With each word of Kuhn, I perceive more clearly the signs that the scientific world needs a new theory. One of these symptoms was published in Scientific American Brasil, number 134, July 2013, the cover story of which is an article by Hans Christian von Baeyer on the quantum model known as QBism, according to which one should not attribute an objective reality to functions wave of

---

[20] https://www.ghtc.usp.br/server/pdf/Sci-Am-eter-2.PDF

[21] KUHN, 2011, pp. 107-123

Quantum Mechanics, but merely the character of a mathematical tool to help identify reality, even denying the possibility of superposition of quantum states, ironized by Schrödinger, with the metaphor of his famous cat that is both alive and dead.

The real symptom is that this is one of many explanatory philosophical configurations of the quantum model, including all versions of String Theory, and also the model that attributes a holographic nature to physical reality. According to Thomas Kuhn:

> *... the crisis, by provoking the proliferation of versions of the paradigm, weakens the rules for solving the puzzles of normal science, in such a way that it ends up allowing the emergence of a new paradigm.*[22]

By denying the physical nature of probability waves, QBism, in this respect, becomes perfectly compatible with AMT. What is most notable, and which coincides precisely with Kuhn's observations, is that the proponents of QBism still cannot break with the quantum uncertainty paradigm, because as long as there is no substitute paradigm, scientists who deny the current paradigms fail to be considered scientists, and this is a phenomenon typical of the evolution of normal science. According to the aforementioned article by HC von Baeyer:

> *... proponents of QBism embrace the notion that, until an experiment is performed, the result simply does not exist. Before an electron's velocity or position is measured, for example, the electron simply has no velocity or position. The fact of measuring attributes existence to these properties.*[23]

This notion is proof of the non-rupture with the uncertainty principle in quantum scales. Certainly, Algorithmic Theory proposes a break with this notion, notably with the notion of collapse of the wave function, but the most interesting thing is to observe how extraordinary are Kuhn's descriptions of the role of paradigms in the evolution of science, especially physics.

---

[22] KUHN, 2011, p. 110

[23] BAEYER, 2013, p. 43

Returning to talk about Maxwell, despite his efforts to engender an electromagnetic ether, we can see, in Kuhn, a detail that may explain, from an algorithmic perspective, the limits to which the conceptions of Maxwell's time were exposed and which may have contributed to that such efforts were unsuccessful. Let's see the following excerpt, very enlightening, referring to episodes that occurred in the 19th and early 20th centuries and that would culminate in the collapse of the ether at that historical moment:

During the middle decades of the century, Fresnel, Stokes, and others devised numerous articulations of the ether theory designed to explain the failure to observe displacement. All these articulations presupposed that a moving body drags some fractions of ether along with it. Each of these articulations succeeded in explaining not only the negative results of celestial observation but also those of terrestrial experiments, including the famous experiment of Michelson and Morley. There was still no conflict, except between the various joints. In the absence of relevant experimental techniques, this conflict never deepened.

The situation changed only with the gradual acceptance of Maxwell's electromagnetic theory in the last two decades of the 19th century. Maxwell himself was a Newtonian who believed that light and electromagnetism in general were due to variable displacements of the particles of a mechanical ether. His early versions of the theory of electricity and magnetism made express use of the hypothetical properties he attributed to this medium. These properties were dropped from the final version, but Maxwell continued to believe that his electromagnetic theory was compatible with some articulation of Newton's mechanical conception. Developing a proper articulation became a challenge for Maxwell and his successors. However, in practice, as had happened many times in the course of scientific development, the necessary articulation proved immensely difficult to produce. In the same way that

Copernicus' astronomical proposal (despite its author's optimism) generated an increasing crisis in the existing theories of motion, Maxwell's theory, despite its Newtonian origin, ended up producing a crisis in the paradigm from which it had emerged. Furthermore, the crisis became more acute of the problems we have just considered, namely, those relating to motion in the ether.

Maxwell's discussion related to the electromagnetic behavior of moving bodies did not refer to the resistance of the ether and made it very difficult to introduce such a notion into his theory. As a result, a whole series of earlier observations designed to detect displacement through the ether turned out to be anomalous. As a result, the years after 1890 witnessed a long series of attempts, both experimental and theoretical, to detect ether-related motion and introduce the latter into Maxwell's theory. In general, the first attempts were unsuccessful, although some analysts considered their results equivocal. Theoretical efforts have produced several promising starting points, most notably those of Lorentz and Fitzgerald, but these also have brought to light new puzzles. The end result was precisely the proliferation of theories that we have shown to be concomitant with crises. It was in this historical context that, in 1905, Einstein's special theory of relativity emerged.[24]

As we proposed at the beginning of this book, that historical moment seems to hold the key to the search for a new paradigm capable of making physics advances. The limitations in the conceptions of the ether, which I referred to earlier, were related to the pre-Einsteinian mechanical paradigm. From Descartes onwards, the ether consisted of microscopic corpuscles. This was the ether with which Maxwell was obliged to work since the paradigm of the corpuscular ether governed a good part of the conceptions of his time. Kuhn asserts that... *"after 1630 and especially after the appearance of the immensely influential works of*

---

[24] KUHN, 2011, pp. 101-103

Descartes, most physicists started from the assumption that the universe was composed of microscopic corpuscles and that all natural phenomena could be explained in terms of corpuscular shape, size, motion, and interaction".[25]

Corpuscular ether vs stratified algorithmic ether

Before having access to the famous book by Thomas S. Kuhn, I had already read something about Descartes' planetary ether. Only now I do realize how comprehensive his conception of the ether was. It would even be worth asking to what extent that conception differs from the algorithmic conception we are suggesting. We are talking about the nature of the particles that make up the ether, which, according to the algorithmic model we are announcing, are composed by the movement of other particles of a lower dimensional scale, as we have already explained to exhaustion: an ether stratified in successive dimensional scales that harbor particles formed by natural patterns of cyclic motion of the particles of the scale immediately below. In a sense, a recursive ether. Not in the exact sense because it does not presuppose forming particles identical to the formed particles. The principle is not entirely recursive because it does not define the shape of the particles. However, it is recursive in the sense that even the fundamental particles of a certain scale are formed by the continuous movement of particles of the scale immediately below. We assume particles of the most varied shapes, sizes and complexities in each dimensional scale.

Perhaps the names adopted so far for the Algorithmic Theory and for the algorithmic principle itself can perfectly represent Cartesian physics. Descartes did not believe in the atomization of the ether, as he argued that each of its particles could be indefinitely broken or worn away. But this idea does not seem to reflect recursion, because even divisible, breakable, or precisely because of this, the particles of the Cartesian ether were solid corpuscles and the great French philosopher did not define the composition or the internal structure of these particles, limiting himself to saying that they arose at the time of creation, originally with irregular shapes, so that there was no empty space between them and which were placed by God in eternal movement,

---

[25] KUHN, 2011, pp. 64-65

maintaining the absence of any vacuum, despite the incessant relative movement between the particles, which causes its fragmentation and wear. Obviously, I cannot agree with this particular aspect of the Cartesian model, as it is incompatible with my algorithmic proposal. Yet Descartes created his own algorithmic principle, his algorithmic matter, and his algorithmic ether, the motion of which determined all physical phenomena. For this reason, perhaps the best way to designate our current model is to incorporate the term "recursive" into the name of the fundamental principle we are proposing, something like the *recursive* algorithmic principle. However, for purely practical and aesthetic imperatives, establishing the necessary differences about the Cartesian algorithmic principle, I prefer to keep more synthetic expressions like "algorithmic principle" and "Algorithmic Matter Theory" (or "Algorithmic Theory") to designate the model of reality that I endeavor to describe.

If this model of algorithmic ether (stratified, recursive) has already been conceived before, the Algorithmic Matter Theory would not be unprecedented and what we are doing here would only be articulating this conception with current scientific knowledge, incorporating the relativity of standards of time and distance[26], returning to space the role of inert stage imagined by Newton. At the beginning of this essay, I stated that I thought that sooner or later I would find historical references to the algorithmic conception that I developed, and then I came to believe, motivated by successive readings, that this conception could be unprecedented. The most comprehensive prior conception I have encountered so far is this reference to Descartes by Thomas Kuhn. Certainly, the algorithmic conception was not considered by normal science at the end of the 19th century, because otherwise there would be no attempt to prove the existence of the ether through experiments that aimed to demonstrate that this medium could cause resistance to the movement of bodies or be simply dragged for this movement. As far as we can see, at that time ether was no longer associated with the constitution of matter. Apparently, the ether was thought of as something

---

[26] About the relativity of distance measurements, we understand that the variation in kinetic density that produces the gravitational effect, as well as the increase in body speed, can not only change the temporal rhythm of the particles' algorithms, but also the size of their structure and their relative distances.

that did not contribute to the existence of material bodies. The maximum interaction between ether and matter was in the sense that the ether could resist the movement of matter or be dragged along by that movement, but not in the sense that matter was the movement of the ether. The eventual drag that the movement of bodies could cause in the ether indicates that matter and ether were substances that were not part of each other's constitution, contrary to what Descartes had already proposed and what we are proposing again now, in different terms.

The new puzzles that arose from the theoretical efforts aimed at improving the concepts of the ether could not be solved at the end of the 19th century. The Michelson and Morley experiment gave rise to the extraordinary conception of the compensation mechanism by Fitzgerald and Lorentz, imagined justifying the ether, and which ended up becoming fundamental for Special Relativity, but contributing to lead Einstein to declare the non-existence of the particulate cosmic medium.

The Theory of Relativity removed the ether from the equation, which allowed the advancement of physics, along the lines described by Thomas Kuhn. But Einstein did not prove the non-existence of the ether, as observed by Roberto de Andrade Martins. The German master demonstrated with extreme competence that the existence of a strictly mechanical ether, formed by corpuscular particles, would be something extremely improbable, from a logical point of view, considering the results of Michelson and Morley. This demonstration does not, however, rule out the recursive algorithmic ether in the sense we are proposing. After showing that the classical transformations are not compatible with Maxwell's equations and with the premise of the constancy of the speed of light, regardless of the speed of the observer, Einstein concluded that this constancy would be necessarily related to the hypothesis that the rhythm of time and distance measurements vary as a function of the observer's speed, adapting Fitzgerald and Lorentz's proposition, but without the ethereal medium. After that, Einstein did not bother to design a type of non-corpuscular ether that would support the relativity of time and distance measurements. I suppose it is reasonable to infer that Einstein was not in a position, concerning his historical interest, to follow the path of seeking a new ether, since his conclusions already represented an extremely dramatic rupture with the physics then in force. From a retroactive perspective, it is not difficult to see that the paradigm that

Einstein was trying to consolidate already had the Herculean mission of supplanting the corpuscular ether and establishing the concept of the relativity of time and distance measurements. To fulfill this objective, its conceptual articulation needed a sufficiently simple aesthetic, since the rival Aristotelian-Cartesian-Newtonian-Huygenesian-Maxwellian articulation, with its evolutionary differences, would not be easily disqualified. The special relativity was urgent, and an ether to explain it could take decades to be created. At that historic moment, science could and should have progressed without an ether. Special Relativity was too important a discovery (as a hypothesis) and deserved to focus all possible attention. It would mobilize the scientific community to devise new tools and methods for dealing with nature. It would change how the scientist should conceive the world and would start a new cycle of challenges to prove the validity of the emerging paradigm. Likewise, it allowed the emergence of General Relativity, without the theoretical inconvenience that the construction of a new ether would certainly cause, which led to the theorization of objects and phenomena not previously considered, such as the existence of black holes and the expansion of the universe.

All this becomes perfectly understandable if we consider that historical-scientific-cultural context, even more so if we keep in mind Kuhn's valuable clarifications on the evolution of the natural sciences. The idea of the relativity of time and space was new, and with Einstein's proposal things seemed to work without the existence of any kind of ether.

In proposing General Relativity to explain gravitation, Einstein needed to follow the paradigm that he had created, and this required an explanation for gravity that also dispensed with the ether. Thus, he proposed the space-time continuum fabric that bends in the presence of matter. The need to explore the logical vulnerability of this concept, especially due to the difficulty of describing its microscopic behavior, has become fundamental for the emergence of our current Algorithmic Theory. Efforts to overcome that gap and to explain the apparent attraction of matter by matter, the attraction of light by matter and the attraction of dark matter by normal matter, as well as the attempt to imagine a way according to which matter could represent a condensed form of energy and maintain a material structure, led me to the conception of the matter formed and sustained by the continuous particles motion in a stratified ether on successive dimensional scales.

Flow mechanisms

We attribute to the differentiated and combined mechanisms of algorithmic flow the explanation for the formation and sustaining of particles of matter, microscopic and macroscopic gravity, the interconvertibility between matter and energy, the phenomenon of inertia, the nature of dark matter, and the phenomena that physics today attributes it to an unknown energy designated as dark energy. We also attribute to algorithmic mechanisms the relativity of time and distance measurements, as well as the compensatory factors that explain the apparent constancy of the speed of light. Furthermore, we propose that this constancy may only valid under local gravitational conditions, and we believe that in extreme circumstances compensation mechanisms such as that by Fitzgerald and Lorentz may eventually have their effectiveness compromised. We claim that our model allows us to speculate that the constancy of the speed of light is related to the number of ether particles consumed, in a given unit of time, by the algorithmic flow involved in the manifestation of the luminous phenomenon, hence the possible locality and possible relativity of the constancy of speed of light. In the field of particle physics, we credit the algorithmic hypothesis the formation of temporary particles obtained by the violent collision of protons in collider artifacts, as well the constitution of force transmission particles, responsible for the cohesion of atomic and molecular structures, and also the emission of all other particles transmitting the forces currently considered fundamental. I have the deepest conviction that the algorithmic model allows generalizing to explain all atomic and subatomic phenomena, including the resistance to gravitational collapse of the particles that make up the structure of the atom, offering a fundamental explanation for all the physical phenomena currently recognized by the scientific community. This includes, of course, the electromagnetic phenomenon. Here we would need to try to envision how to reconcile gravitational and electromagnetic phenomena, a task that consumed so much of Einstein's attention. The task still does not seem easy, but we believe that the math and computational delineation and development of submodels, including competitors, also supported by the algorithmic premise, is the only way.

The Construction of Algorithmic Mechanics

The Algorithmic Matter Theory is not a complete theory of nature, as are classical, relativistic, and quantum mechanics. Rather, it is an eminently qualitative new vision of how the universe can function. It does not answer all the questions, but it does create a new class of them, as one would expect from any candidate theory for a new universal paradigm in physics. It invites a deep imaginative dive into the structure of matter and challenges science to start a path to progressively establish the dynamic architectures of nature's main algorithms, evolving little by little from the most rudimentary representations to more consistent and refined forms. We do not believe it will be a smooth journey. The previous models, although in crisis, are very sedimented in the minds of scientists. Theoretical physicists realize that the problem is deep and they strive to create, with increasing profusion, alternative models derived from the fundamental assumptions still accepted today. However, perhaps very few have already realized that it would probably be more productive to go back some important steps so that we can move forward, radically breaking with some largely hegemonic assumptions. As long as the uncertainty principle and the concept of space-time remain untouched, despite some notable divergent point efforts, I claim that the problem of current physics will not be solved. As the algorithmic premise is tested, looking at the experimental data available and the anomalies already identified from this new perspective, I have no doubt that it will be possible to design increasingly efficient means for the creation and development of algorithmic mechanics. Undoubtedly, any initiative in this direction will have to use sophisticated computer programs to simulate the possible algorithmic mechanisms, as is done, for example, to simulate the dynamics of the collision of galaxies. Existing knowledge in areas such as fluid mechanics, thermodynamics, astrophysics, particle physics, relativity, and statistics, among others, should be integrated and used as starting points to try to gradually validate the algorithmic premise. Many theorists and experimentalists will indeed feel the need to declare, in the first hour, the falsity of the new model, which we believe to be falsifiable, as Karl Popper's theory demands, but I believe that their arguments will be challenged by an increasing number of researchers, theoreticians and experimentalists, as well as philosophers and historians of science, who will feel challenged to explore this new vision of reality.

Only with the progressive improvement of computing algorithms will it be possible to represent material algorithms in a way that approaches the observable reality, but, before that, a theoretical dialectic may imply a growing success of the algorithmic premise in the task of explaining the most varied experimental results obtained and current controversial phenomena.

This optimism comes from intuition and confidence in the model we are proposing, in the wake of more than ten years of research and reflection with a specific focus on validating or falsifying the model, we are presenting.

It is possible to clearly perceive a growing tendency to redesign the ether, both in "normal" science and in "alternative" science, but what we have seen is an effort to create an increasingly quantum ether, which tries to explain the functioning microscopic view of the space-time fabric. I believe that in the current historical context, Ockham's razor favors the abandonment of space-time and the uncertainty principle, in its conceptual formulations, and its corollaries, just as another scenario once favored the abandonment of the ether in favor of relativity. The stratified algorithmic ether is a much simpler explanation, with an explanatory potential of a broader range, eventually unlimited.

The challenges of the algorithmic premise

To function as a consistent model, the theory must provide a logical answer to any physical problem, even if it is a generic answer, but it must always strive to imagine a solution to each enigma from an algorithmic perspective, sketching possible flow architectures, integrating the observation and existing knowledge about the material world. Generally speaking, it is obvious that any problem presented to the model must be answered by algorithmic mechanisms that process the particles of sustaining matter attracted by the material algorithms. Its mathematization will probably not be easy, as well as the construction of computational models that simulate reality from the perspective of the theory.

CHAPTER VI

OTHER ELEMENTS UNDER THE ALGORITHMIC PERSPECTIVE

Concepts, considerations, and experimental results

In an effort to provide explanations compatible with the new theory, we began to outline concepts, make new considerations, and discuss, from the algorithmic perspective, important experimental results, some of which have not yet been addressed in this book.

Energy

The measure of the amount of movement. All energy is kinetic at some dimensional level. Different algorithms may be subject to different limits on energy they can absorb without changing the patterns of their motion architectures. In addition to quantitative limits and general environmental conditions, the architecture stability of a given algorithm may also be subject to morphological aspects of the flow system in which it participates. The architecture of the considered algorithm certainly influences this aspect, including concerning linking algorithms, which only form when there are relatively stable groupings of simpler algorithms. They will also be relevant, for the stability of the material body and the particles of matter or energy of our human scale, variables related to the local conditions of the algorithmic medium, particularly the kinetic concentration, average chaotic agitation speed and average size of the first order particles, variables strongly influenced by macroscopic gravitation.

Mass

The measure of the ability of a particle or body algorithm to attract sustaining particles to its material algorithm, increasing the kinetic concentration of the medium in its surroundings.

Microscopic gravity

The flow of sustaining particles and kinetic turbulence induced by the presence of an elementary particle of a certain dimensional scale.

Macroscopic gravity

The measure of the kinetic concentration of particles of the ethereal medium induced by the presence of a material object or set of material objects of a certain dimensional scale.

Atomic cohesion

It results from the algorithmic interaction between fundamental particles in our dimensional scale. This interaction creates a composite motion algorithm, with sub-flows and specific particles that can exist only when this interaction occurs.

Magnetic field

Bodies or material particles form flow patterns in the sustaining medium, being able to induce, under certain conditions, an ordering of the medium itself in movement patterns, transforming the chaotic movement of its particles into more or less ordered flow patterns, but without constituting material particles. The comparison can be made with the static wave patterns that a stream forms around a stone, or with milk poured from a pitcher before it reaches the cup. These patterns influence the motion of certain kinds of particles or material bodies. While gravity can be seen as a gradient of kinetic chaos, the magnetic field can eventually be interpreted as a gradient of order in the kinetic sustaining medium.

Electric charge

A special algorithmic property of certain material particles that causes a certain kinetic reaction in the movement of other particles with special characteristics, being the algorithmic medium the messenger of this reaction.

Electric field

The movement patterns in the sustaining medium induced by the algorithm of certain types of fundamental particles of our dimensional scale and which in turn influence the movement of special, similar or different particles, sensitive to the ordered patterns of the algorithmic environment.

Electromagnetic waves (light)

I am not capable of totally discarding the wave character of what is currently known as an electromagnetic wave, at least until satisfactory algorithmic models for light are built with mathematical and computational tools. However, I tend to assume that light is formed by algorithmic particles. In one case or another, it is a disturbance of an algorithmic nature in the sustaining medium. Possibly, certain wave characteristics attributed to the so-called electromagnetic waves are due to particle shots at regular intervals forming circular beams from the emission source.

Photons

In our algorithmic view of the photon as a particle, we believe that it is a "portion" of movement that strays from the electron algorithm, and can also be temporarily absorbed by an electron in the atomic structure.

Newton's bucket

We answer the question of how Newton's spinning bucket works by stating that it rotates in relation to the particulate medium, which acts as a privileged frame of reference. The water rises near the wall of the bucket due to the algorithmic inertia of the water particles. The bucket and the rotating water form a system that follows the laws of flow, dictated by local flow conditions. The role of fixed stars and distant galaxies is only that of being part of the structure of the cosmos and contributing to a certain homogeneity of cosmic functioning.

Double-slit

One of the explanations that derive from the Algorithmic Theory to the results of the double-slit experiment is that the particles of the plate material that contains the slits mobilize the ether because of the movement algorithms that form these particles. When the plate has a single slit, material mobilization forms a standard of motion in the empty region where the slit is located and surroundings. This mobilization of the ether around the slit, forming a standard, even explains the diffraction of light in a single slit. When there are two nearby slits, the ether mobilization patterns of each slit interfere with each other, forming regions with aspect of static waves, such as the mobilization of stream water as it passes between two stones side by side. It is this new pattern, formed by interference between the mobilization patterns of the ether in each slit region, which determines the trajectories of the light or matter particles launched against the pair of slits and the probability of these particles of light or matter reach the bulkhead to form the fringe pattern. In short, one slit interferes with the other.

Another possible explanation, based on the algorithmic premise, is that the particle of light or matter creates a disturbance in the ethereal environment as it moves, and this disturbance crosses the two slits at the same time, before the arrival of the particle. When crossing the slits, the disturbance forms an interference pattern from the slits, which influences the movement of the particle and determines the probability that the particle reaches a specific location in the experiment's bulkhead.

These two alternative explanations come near the idea of Broglie-Bohm pilot wave, although the mechanisms involved are distinct. In the algorithmic mechanism, for example, there are no wave functions such as physical entities, nor considerations of non-locality, hidden variables are not necessary, and the disturbance that accompanies the test particles is not a disturbance necessarily in wave format, before it reaches the pair of slits.

Lorentz contraction

It is not ruled out by our theory, however, if it is a reality, it has an algorithmic nature.

Curvature of space

As we have claimed, space has no curvature and there is no continuous space-time as a physical reality. What exists are mathematical curves corresponding to the kinetic concentration gradients of the sustaining medium.

Spatial folds

If they are possible, they will not have the nature of space-time folds but will consist of some kind of manipulation of the flow of dark matter (sustaining matter).

Existence of an absolute vacuum

According to our algorithmic perspective, there is no such thing as an absolute vacuum, considering the different dimensional scales of the cosmic environment. On the other hand, the vacuum relative to fundamental particles of matter and energy of a certain dimensional scale is perfectly possible.

Probability waves

Mathematical objects that do not capture the nature of particles of matter and energy in our perspective. However, the mathematics of current particle physics is so powerful that there must be a relationship between this mathematics and the movement algorithms that characterize the fundamental particles of our dimensional scale.

Uncertainty principle

It is a mathematical principle that, although very vigorous, does not make explicit the nature of particles of matter and energy, according to our algorithmic perspective.

Multiple paths taken by quantum particles

Explanation for the results of the double-slit experiment that is not related to the behavior of nature.

Quantum foam

Unnecessary concept from an algorithmic perspective.

Large Hadron Collider

The images produced by the detection techniques of the Large Hadron Collider seem to impressively illustrate the algorithmic nature of material phenomena.

Falling bodies and acceleration due to gravity

Imagine a body formed by several algorithmic flow circuits (matter particles). These circuits are connected by a set of flows, no matter what form they take, which are linking flows. If the reader has ever seen a video of a large tornado surrounded by smaller tornadoes dancing around it, it's a fair comparison. Everyone will agree that there is a connection between the big tornado and its little offspring. This link is algorithmic, a flow link. Imagine that, by hypothesis, a cyclone has its path determined exclusively by the ocean temperature factor. Imagine that the cyclone is attracted by regions of the ocean where the water temperature is higher, due to the dynamics of sustaining the flows that form it. Analogously, a material body is attracted by gravitational regions that are "warmer", that is, that have a higher kinetic concentration of the support particles. The regions of the higher kinetic density of sustaining particles exert an attractive effect on material particles (and bodies). Gravity forms a decreasing kinetic concentration gradient of first-order particles around bodies with mass – the gradient decreases as we move away from a body.[27] Since the gradient increases as we move close to the body, and its center, the nearby bodies accelerate to each other, in the measures calculated by Newton's and Einstein's equations. The algorithmic system of the body, such as the tornado system cited as an example, is attractively sensitive to the kinetic concentration gradient caused by the other body, and vice

---

[27] In the case of particles with mass, the very algorithm that constitutes it exerts, under certain conditions, an attractive effect for other particles with mass. The flow conditions that determine the charge of the particle and other properties govern the final behavior of one particle relative to another.

versa, because the increase in the kinetic concentration of the algorithmic sustaining medium _favors_ the feeding of the material flow of each algorithmic particle of the body and simply the particles of the body are _sucked_ by the environment of greater kinetic concentration. In this scheme, we make the necessary bridge between the microgravity of the particles of a body – with the increase in the kinetic concentration that this microgravity causes – and the external macro gravity of the body, in a physical way, without the drawbacks of trying to establish a physical interaction between a geometric substrate and statistical particles of a material object.

Shearing of a fluid

If we can see the cosmos as a great algorithmic fluid that crosses and constitutes matter while being permeated by it, in all dimensional scales, in an infinite system of particles formed, successively, by particle flows, some specific properties of this fluid are yet to be discovered, and other features of material fluids known to us may eventually be shared. One is the shearing of a fluid, as happens, for example, with atmospheric air and water currents. At the boundary between air or water currents, two medium currents move in opposite directions – or at different speeds, depending on the frame of reference – in a phenomenon called the shear of the medium.

Gravity, dark matter and shear

Objects with mass are the result of algorithmic flows of dark matter – sustaining matter – from an ether stratified into infinite dimensional layers. All matter particles, at all scales, are formed by some kind of flow, architecture of which remains relatively stable while the dark matter particles of the next lower scale enter and leave the flow algorithm, engulfed and deposited somewhere in the vicinity of the considered algorithm, which probably, as a rule, is in a translational movement in the medium that sustains it. Each particle of matter, when swallowing and returning to the sustaining medium the particles that participate in the material algorithm, shakes the particles of the ethereal medium. Indeed, they already cause an uproar in the first-order ethereal medium since their translational approach through that medium and return the sustaining particles, in an uproar, to the surroundings of the material algorithm. All this turbulence also affects the particles of the underlying scales that participate in the algorithms of first-order particles, agitating the medium at least a few scales down, until the equilibrium point is reached with the natural agitation present in a certain lower scale. This kinetic agitation of the first order and underlying particles forms a decreasing kinetic energy gradient from the material particle under consideration. In other words, the further away from the particle, the less agitated the sustaining medium is. The algorithmic (ordered) movement of the material particle carries with it, wherever it goes, the chaotic, turbulent motion of first-order and underlying particles. The order carries its chaos, in reciprocal feeding. Heraclitus would love that! Here we need to make a distinction. The chaos caused by a particle in translational motion does not seem to account for its microscopic gravity any more than its material algorithm. In a body, the chaotic kinetic concentration caused by its countless particles causes the body's gravity, as we claim. Let's move from particles of matter to objects of great mass like the Earth, the Moon and the Sun, or the other stars of our galaxy, formed by billions of trillions (numerical expression chosen randomly) of particles of matter. These objects create a large kinetic chaos around them that attracts other objects with mass wandering nearby. The material pressure inside the massive object increases as its center approaches. This increases the number of material particles per unit volume, which means that it increases the concentration of matter particle algorithms inside the body. The chaos created by objects like these, of the order of magnitude of planets or stars, in the surroundings of the body, does not necessarily have a well-defined border, and it gradually decreases, as the distance from the material object increases, until the gravitational chaos balance with the natural chaos inherent in the medium formed by first-order particles (and underlying orders), in areas with negligible gravity. We call this energetic chaos macroscopic gravity, in our dimensional scale, the human scale. In large groupings of material bodies, such as galaxies, a flux shear may occur, or more than one, since the particles of the fluid medium, so concentrated, can cause the formation of current boundaries, regions of relative displacement between the entire grouping (the entire galaxy) and a more external region of the sustaining medium, or between concentric layers of the grouping, forming one or more rotating currents with different angular velocities from the center of the system, which break away from the outermost medium. This, certainly, explains the effect under which stars on the periphery of some galaxies do not present the angular velocity expected according to Newtonian equations or General Relativity. The problem is that such mathematical formulations, as well as the most powerful telescopes, do not capture the shear effect of the cosmic fluid and consider individually each object and its reciprocal effects, without taking into account the shear limit of the sustaining medium eventually exceeded due to the immense concentration of matter.

BIBLIOGRAPHICAL REFERENCES AND FURTHER READING

BAEYER, Hans Christian von. O Paradoxo Quântico que Confunde a sua Mente. In Scientific American Brasil, Ano 11, N° 134 – julho de 2013.

DESCARTES, René. Princípios de Filosofia. Lisboa: Edições 70, 2006.

EINSTEIN, Albert; INFELD, Leopold. A Evolução da Física. Rio de Janeiro: Zahar, 2008.

EINSTEIN, Albert. Como Vejo o Mundo. Rio de Janeiro: Nova Fronteira, 1981.

GREENE, Brian. O Universo Elegante. São Paulo: Companhia das Letras, 2001.

HAWKING, Stephen W.; MLODINOW, Leonard. O Grande Projeto. Rio de Janeiro: Nova Fronteira, 2011.

JACOBSON, Theodore A.; PARENTANI, Renaud. Propagação nos Buracos Negros. In Scientific American Brasil, Ano 4, N° 44 – janeiro de 2006.

KUHN, Thomas S. A Estrutura das Revoluções Científicas. São Paulo: Perspectiva, 2011.

NOVELLO, Mário. Os Cientistas da Minha Formação. São Paulo: Editora Livraria da Física, 2016.

PATY, Michel. Introdução a três textos de Einstein sobre a geometria, a teoria física e a experiência. São Paulo: Scientiæ Studia, 2005.